Learning Materials in Biosciences

Learning Materials in Biosciences textbooks compactly and concisely discuss a specific biological, biomedical, biochemical, bioengineering or cell biologic topic. The textbooks in this series are based on lectures for upper-level undergraduates, master's and graduate students, presented and written by authoritative figures in the field at leading universities around the globe.

The titles are organized to guide the reader to a deeper understanding of the concepts covered.

Each textbook provides readers with fundamental insights into the subject and prepares them to independently pursue further thinking and research on the topic. Colored figures, step-by-step protocols and take-home messages offer an accessible approach to learning and understanding.

In addition to being designed to benefit students, Learning Materials textbooks represent a valuable tool for lecturers and teachers, helping them to prepare their own respective coursework.

More information about this series at http://www.springer.com/series/15430

Alberto Pasquarelli

Biosensors and Biochips

 Springer

Alberto Pasquarelli
Institute of Electron Devices and Circuits;
Faculty of Engineering, Computer Science
and Psychology
Ulm University
Ulm, Germany

ISSN 2509-6125   ISSN 2509-6133   (electronic)
Learning Materials in Biosciences
ISBN 978-3-030-76471-5   ISBN 978-3-030-76469-2   (eBook)
https://doi.org/10.1007/978-3-030-76469-2

This Springer imprint is published by the registered company Springer Nature Switzerland AG.
The registered company address is: Gewerbestrasse 11, 6330 Cham, Switzerland

To my mother for the past
To my wife for the present
To my children for the future

Preface

There is a rising need for biochemical detection and analysis in many fields, worldwide. Several reasons drive this trend, for instance the dramatic increase in the prevalence of diabetes and the growing awareness regarding environmental control and healthcare, all targeting a better life quality. In response to these demands, new legislative and regulatory standards have been endorsed for many sectors, including, among others, air, water, and soil monitoring, industrial processes, food production and distribution, and drugs development. Thanks to higher sensitivity and specificity, short response times, and reduction of overall costs, biosensors can be very competitive in addressing the mentioned needs and have therefore the potential of not only outperforming traditional methods but also actually enabling for the first time a capillary monitoring, for the benefit of every person in every place.

By combining the biosensing know-how with the power of micro- and nanotechnologies, biochips represent the natural evolution of biosensors offering arraying capability and boosting by order of magnitude the data throughput, thus enabling much faster times for giving robust answers to crucial questions in many fields.

The newest trends in the direction of personalized medicine triggered by the availability of whole-genome sequencing and the recently discovered pluripotent stem cells and their exploitation in the emerging organs-on-a-chip are no longer science fiction but are becoming a very realistic perspective for the near future.

Ulm, Germany   Alberto Pasquarelli

Acknowledgments

I want to express my deep gratitude to all colleagues of the School of Advanced Professional Studies of Ulm University for their continuous and enthusiastic support, and to all my students who contributed with their comments and suggestions to a progressive improvement of the didactical material, from which this textbook originates.

About the Book

This textbook was written having in mind the goal of fitting a semester course at the master's level. Due to the extent and the broad multidisciplinarity of the matter, spanning from chemistry to computer science, from molecular biology to micro- and nanotechnology, from physics to legislation and regulation, it is not strictly focused on a specific study plan. The book targets students of many scientific areas, while aiming at stimulating with concise explanations the interest of the readers for further learning on a self-guided basis. This is actually in agreement with my teaching experience at Ulm University; in fact, in my classes students from the engineering school, natural sciences, and computer science are all equally welcome, and many of them choose soon after a project for their Master thesis which connects their major with the field of biosensing. Of course, it is necessary to deal with the variety of educational backgrounds in such a pool of students, in order to reach rapidly a common knowledgebase. Doing this I tried to find a balance between providing/recalling essential concepts to fill possible gaps and avoiding to be redundant with excess of basic information, which should be given for granted.

Visual information plays an important role in all chapters. My opinion is that sometimes a good picture can help in grabbing complicated concepts with intuition, more than a long and detailed descriptive text. I hope that the readers will enjoy this presentation method. Moreover, mathematical formalisms are kept to a minimum, limiting their use to the few topics that need a well-defined quantitative description.

I believe that with this approach the book can be appreciated not only by students, but also by all those persons who simply have a genuine interest in understanding how sciences can influence and possibly improve their everyday living experience.

Contents

Introduction

1

Contents

Keywords

Definition · Bioreceptor · Physical transducer · Market potential · Applications

© The Author(s), under exclusive license to Springer Nature Switzerland AG 2021
A. Pasquarelli, *Biosensors and Biochips*, Learning Materials in Biosciences,
https://doi.org/10.1007/978-3-030-76469-2_1

What You Will Learn in This Chapter
In this chapter, you will gain an overview over the functions of biosensors and the history behind their development, as well as various types and their properties. Furthermore, you will learn about the meaning and potential of these sensors alongside their most important usages. You will then be able to explain the advantages upon usage of biosensors, name various application fields, and make statements about the development of biosensors.

1.1 Biosensors

"A biosensor is an analytical device, which converts a biological response into an electrical signal" [1]. This definition of biosensor is very widely accepted; however, some sensors employed for measuring biological parameters are designated as *biosensors*, even if their sensing function is not based on biological receptors.

Bioreceptors are generally very specific and tailored to a particular application (Fig. 1.1).

A biosensor system consists, generally, of the following components (see Fig. 1.2):

(a) Bioreceptor, specific and selective for the analyte to detect
(b) Interfacial component, which binds and protects (a) and (c)
(c) Transducer, delivers a physical output signal like an electrical current (e.g., electrodes).
(d) Analog signal conditioning (amplifier, filter, etc.)
(e) Data conversion, acquisition and processing (if needed)
(f) Display and storage

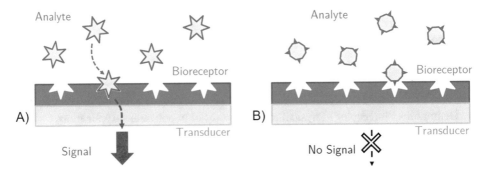

Fig. 1.1 Bioreceptor selectivity: (**a**) If the analyte matches the bioreceptor, a signal is generated. (**b**) On the other hand, a non-matching analyte does not produce a signal at the output

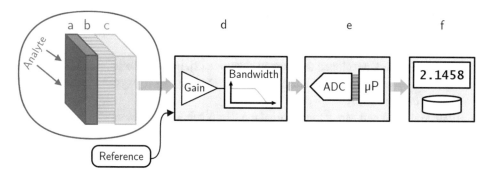

Fig. 1.2 Components of a complete biosensor system. This book focuses primarily on (**a**) the bioreceptor, (**b**) the interfacial component, and (**c**) the physical transducer (circled in red)

Question 1

What is a biosensor? What components typically make up a biosensor system?

1.2 The Development of Biosensors

Biosensors were developed by combining the selectivity of biomolecules with the processing power of microelectronics, providing us with powerful detection and diagnostic tools. Ideally, biosensors should enable better measurements and analysis compared to the traditional analytical methods.

Advantages of biosensors:

- Require smaller sample sizes
- Sensitive, selective, and specific
- Easier to use
- Faster procedures
- Cost effective
- Allow online monitoring

Areas of application:

- Medicine (e.g., clinical diagnostic)
- Pharmaceutical drug analysis
- Industrial processes
- Food quality control
- Environmental monitoring
- Monitoring of biological and chemical threats (terror attacks or warfare agents)

In the following we summarize some of the milestones in the development of biosensors:

- 1953: L. D. Clark developed the oxygen electrode [2]. In 1962, he proposed the enzyme electrode by entrapping the enzyme glucose oxidase at an oxygen electrode using dialysis membrane [3].
- 1969: Guilbault and Montalvo reported the first potentiometric biosensor, a urea sensor based on urease [4].
- 1972/1975: Yellow Springs launched the first commercial glucose analyzer based on Clark's ideas.
- 1974: Mosbach proposed the use of enzymes with thermal transducers for biosensors [5].
- 1975: Diviés suggested the use of bacteria as biosensitive element for measurement of alcohol [6]. Lubbers and Opitz described the first optical biosensor, a functionalized fiber-optic sensor with immobilized indicator (optode) [7]
- 1976: Clemens et al. incorporated a glucose biosensor in a bedside artificial pancreas [8]. Roche introduced the Lactate Analyser LA 640, using hexacyanoferrate as mediator for the electrons transfer.
- 1982: Shichiri et al. described the first needle-type glucose biosensors for subcutaneous implantation [9].
- 1983: Liedberg described real-time monitoring of affinity reactions by "surface plasmon resonance" (SPR) [10].
- 1984: Ferrocene was introduced as an immobilized mediator for use with oxidoreductases in the construction of inexpensive devices, leading to the screen-printed enzyme electrodes launched by MediSense in 1987, for home blood-glucose monitoring. MediSense's sales grew exponentially, reaching US$ 175 million by 1996 when Abbott purchased them. Roche Diagnostics (formerly Boehringer Mannheim) and Bayer now have similar biosensors and these three companies dominate 85% of the world market for biosensors home diagnostics.
- 1990: Pharmacia launched the BIAcore, based on SPR (GE-Healthcare, now Cytiva, acquired BIAcore in 2006).
- 1992: Abbot introduced the i-STAT, portable blood analyzers [11].

Academic journals now contain reports on a myriad of devices, which combine a huge variety of bioreceptors and transducer technologies. In general, the focus is on the development of methods to improve the sensitivity, selectivity, stability, and costs of biosensing technologies.

Question 2
Can you mention some advantages and fields of application for biosensors?

1.3   Biosensor Components

A biosensor is normally made up of three fundamental components:

- Biological component (bioreceptor): enzymes, microorganisms, antibodies, proteins, DNA arrays, chemoreceptors, tissues, or organelles.
- Physical component (transducer/converter): amperometric, potentiometric, semiconductor, thermometric, photometric, or piezoelectric.
- Interfacial component (polymeric thick- or thin-film, chemically modified surface) as connection between the two other components.
  Here we give a first in-depth look at them:

1.3.1   Biological Component

The biological component reacts or interacts with the specific substances to detect. We divide the biological component into two groups: *catalytic* elements and *non-catalytic* elements. Enzymes, microorganisms, and tissues belong to the first group, whereas antibodies, membrane receptors, nucleic acids, and synthetic receptors belong to the second group.

Enzymes are proteins that recognize a substrate molecule and accelerate the chemical reaction for its conversion into the product molecule (acceleration up to 10^{14} times), through catalysis, that is, without influencing the reaction equilibrium. The signal is either measured indirectly, for instance via the change of oxygen concentration, pH, H_2O_2, color, temperature, etc., or through direct electron transfer from the redox enzyme to the physical transducer.

Antibodies are proteins as well. They work as bioreceptors in the so-called *Immunoassays*:

- They are y-shaped immunoglobulins with two identical specific binding sites (Fab's).
- They are produced as natural recognition molecules by the immune system of a body.
- They are able to bind several antigens (viruses, microbes).
- The signal output is optical, gravimetric, impedimetric, etc.

Question 3

What is the function of the biological component of a biosensor?
    Which categorization do you know for this component?

1.3.2   Physical Component

The physical component converts the biochemical reaction or interaction into a quantifiable and processable electrical signal. This can be a simple electrochemical device or a complex, multi-technological instrument. The signal conversion can be based on various methods:

- Amperometric
- Potentiometric
- pH-metric
- Calorimetric
- Thermal
- Optical
- Gravimetric
- Others

1.3.3   Interfacial Component

The interfacial component links the biological component with the transducer. Its primary function is to immobilize the bioreceptor. Furthermore, it can have a selective and/or protective function (e.g., a thin film membrane) for the cases of labile, toxic, or corrosive substances, or even for preventing interferences and other sources of artifacts and measurement errors. Examples are membranes, conductive polymers, gel matrices, or self-assembled monolayer (SAM) of cross-linkers.

The following figure shows the structure of a traditional glucose biosensor as an example of the interconnected functions of the three components in a biosensor (Fig. 1.3).

Question 4
Which methods do you know for signal conversion of the physical components?

1.4   Biosensors Significance and Its Market Potential

Biosensors have the potential to influence many areas. The main fields of application include medicine, environmental monitoring, food quality control, and control of industrial processes, but also physical therapy, and even entertainments like music or video games can all potentially benefit from the introduction of biosensors.

Blood glucose sensing for diabetics plays a fundamental role in supporting biosensor research, development, and their successful commercialization. Diabetes monitoring products represent by far the largest biosensor market, because no other disease combines the two requirements for mass production: (1) The incidence on a large population

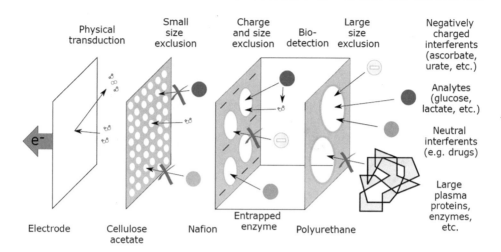

Fig. 1.3 Structure of a "classical" glucose biosensor

(as reported by the World Health Organization, the global prevalence of diabetes among adults over 18 years of age rose from 4.7% in 1980 to 8.5% in 2014 [12]); and (2) the necessity for frequent measurements (on average two to four times per day) of clinical data (i.e., blood glucose concentration).

For comparison, a patient affected by a cholesterol disorder requires only one measurement per week or even per month, which significantly diminishes the required production volume, even though a large proportion of the population needs this kind of monitoring.

The progress of biosensors over the last 50 years begun with the development of the biosensor concept as a simple, specific, robust, cost-effective, portable, and easy-to-use technology. However, throughout the whole first generation of such devices, the focus was mostly placed on integration, thus realizing sensor systems with the combination of multiple steps or analysis of multiple analytes. This was expensive, while requiring both trained users and dedicated laboratory environments.

In the second generation, the focus was moved on miniaturization, which enabled smaller integrated systems: mass production and cheaper components became a reality.

Despite those progresses, the lack of equivalent mass markets beside blood-glucose monitoring, discourages investment in diagnostic biosensor technologies with a view to commercial profit. Nevertheless, due to the real and potential advantages of biosensors compared to more traditional methods of analysis, many market opportunities do exist in the areas of medical diagnosis, environmental monitoring and analysis of food and beverage quality, industrial and pharmaceutical products and processes (Fig. 1.4).

Several key-parameters influence the development of the biosensor market. They can be classified in six groups as shown in Fig. 1.5, which depicts them in vector diagram.

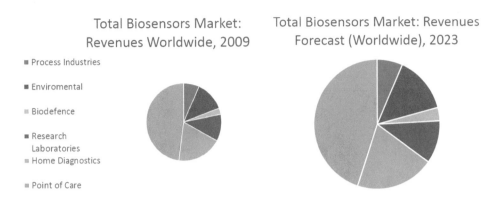

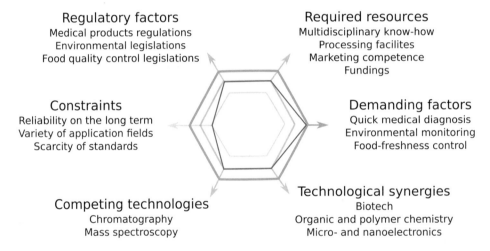

Fig. 1.4 The market segmentation for biosensors between 2009 and 2023 is predicted to remain largely unchanged. The total market volume is expected to triplicate

Fig. 1.5 Six groups of parameters influence, as force vectors, the commercial development of biosensors. For a successful strategy, the global profile (red) should be as regular as possible

In the following we give closer look only at three of these groups, namely constraints, demanding factors, and technological synergies.

1.4.1   Constraints

The most relevant constraints and their influences on the development are as follows:

• Low durability and short shelf life lead to expensive single-use devices.

- Fear of a rapid decline in profits if biosensors became commodities, causing manufacturers to be reluctant to make manufacturing commitments.
- Since there are currently no "standard" multi-function biosensors for a wide range of users, this leads to concentration on markets with high demand potential, for example for blood glucose monitoring.
- So far, most financial resources have been invested in the development of the technology and the marketing remains neglected. Therefore, profitable products are in short supply.
- The research and development of biosensors require a variety of competences (semiconductors, electrochemistry, optics, polymer chemistry, microbiology, biochemistry, etc.). Therefore, only a few companies or institutions have the resources to put together such multidisciplinary teams.

1.4.2   Demanding Factors

On the other hand, development is driven by demand. The biggest demand is healthcare. Drug development, pollution, and food analysis are other important demanding areas.
    Some of the most relevant applications in the respective sectors are:

- Healthcare: Rapid spread of diabetes, rapid monitoring of high-risk medicine concerning side effects, rapid diagnosis for the seriously ill, rapid in situ analysis of critical parameters.
- Drug testing: Preclinical toxicity testing, monitoring during clinical trials, and quality control in drug production.
- Pollution: Increasing need for environmental monitoring, facing the threat of ABC (atomic, biological, chemical) terrorism.
- Food analysis: New legal standards for microbiological monitoring.

1.4.3   Technological Synergies

The convergence of multiple technologies behind biosensors is very beneficial. The three most relevant disciplines are biotechnology, polymer chemistry, and microelectronics. An example to explain this concept in a better way is represented by the deposition of metal layers at low temperatures onto polymer substrates instead of depositing the same metal onto expensive ceramic materials. Another example is the computer-aided design (CAD) of enzymes and biomimetic molecules, for example, aptamers, which greatly reduces the need for expensive monoclonal antibodies, recombinant DNA, and other rare substances.

Question 5
Which factors foster the development of biosensors? Which factors are more of a hindrance?

1.5   Applications

The following survey presents some relevant examples of necessities in the various fields mentioned above, which are addressable with biosensing technologies.

1.5.1   Healthcare

Applications in this area are:

- Clinical blood analysis, point-of-care systems: O_2, CO_2, pH, glucose, K, Na, Ca, Cl, bicarbonate, urea, creatinine, lactate, hematocrit, bilirubin, and cholesterol.
- Cancer diagnosis based on the analysis of urine samples.
- DNA diagnosis for the discovery of genetic diseases.
- Bedside monitoring in intensive care.
- Kits for self-diagnostics at home, like blood glucose, ovulation/pregnancy, cholesterol, urea, sport and fitness monitoring.
- Veterinary diagnostics for domestic animals like pets and cattle.

Rapid diagnostics is one of the advantages of biosensors. Since biosensors deliver results almost instantly, they allow performing clinical diagnoses at the bedside in intensive-care units, greatly reducing the need for lengthy laboratory procedures to be carried out in centralized laboratories. The treatment of patients can therefore often start immediately.

Furthermore, consumer-oriented rapid diagnostic products make self-testing at home possible, thanks to the adoption of user-friendly formats and extremely simplified protocols.

The common characteristics of all these products are small sample volumes, faster diffusion of reactants, reduction of background noise, easier filling and manipulation, as well as affordability and portability. In particular, consumer products ("low tech" approach, e.g., pregnancy tests) include direct immersion, sample drops onto sensor, lateral flow (*wicking*) and capillary filling (Fig. 1.6), while more sophisticated approaches like microfluidic "lab-on-a-chip" devices are professional clinical products (Fig. 1.7).

Fig. 1.6 Examples of consumer products. Left: Pregnancy test as consumer product in "low tech" approach (Source [13] Klaus Hoffmeier, public domain). Center: Urine dipsticks for rapid screening at home of three parameters. Two unreacted strips (above) and one strip (below) reacted with a sample having pH 7.2, 0.3 g/l proteins, and 5 mmol/l glucose. Right: Legend for reading out the test results (disclaimer: the color scale is shown solely for illustrating the concept, it is not intended as reference for any real diagnostic purpose)

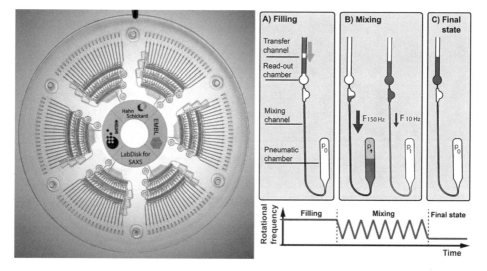

Fig. 1.7 Example of a professional lab-on-a-chip. Centrifugal microfluidic LabDisk for protein structure analysis via small-angle X-ray scattering (SAXS) on synchrotron beamlines. (Left) The device prepares 120 different measurement conditions, grouped into six dilution matrices. (Right) Mixing via reciprocation. Aliquots are combined in the read-out chamber (**a**). By the alternation of rotational frequencies between high and low frequencies, the liquid is pushed back and forth between the read-out chamber and the pneumatic chamber (**b**). After 10 cycles mixing is completed and the dilution matrix can be transferred to the beamline for read-out (**c**). (Reprinted from [14], © The Royal Society of Chemistry 2016, CC-BY 3.0 [15])

For rapid screening at home (consumer products), it is possible to measure a range of common variables from a urine or saliva sample. Further advantages of these formats are as follows:

- The results are available in just 3–5 min.
- It is a simple process performed in a single step.

Fig. 1.8 i-STAT
200 (Reprinted by permission
from Springer Nature [18],
© 1997)

- It provides a certified laboratory-grade accuracy.
- Sample handling is not necessary.
- There is no need for qualified clinical operators.
- Extra reagents are not required.
- Pipetting is not necessary.
- Quality control is built-in, for instance as a change in color.
- The shelf life is usually very long, spanning from several months to a few years.

A milestone in the field of professional products was the handheld *i-STAT 200* by Abbott as portable blood analyzers, allowing several parameters to be measured directly and a few more calculated, delivering the results in just a few minutes. In more recent years, this sensing technology evolved in new products like the *i-STAT 1* and the *i-STAT Alinity* both by Abbott [16], and the *epoc* by Siemens-Healthineers [17] (Fig. 1.8 and Table 1.1).

1.5.2 Environmental Monitoring

Compliance with environmental regulations can be assessed by means of biosensors. Measuring the Biological Oxygen Demand (BOD) of wastewater is one of the most

Table 1.1 Direct and indirect parameters of the i-STAT 200 (Source: Abbott Point of Care Inc.)	Measurable parameters	Calculable parameters
	Sodium	Hemoglobin
	Potassium	Total CO_2
	Chloride	Bicarbonate
	Glucose	Base excess
	Lactate	Anion gap
	Creatinine	O_2 saturation
	pH	
	PCO_2	
	Ionized calcium	
	Urea nitrogen	
	Hematocrit	
	Celite ACT	
	Kaolin ACT	
	Prothrombin time	
	Troponin I	
	Creatine kinase MB	

important tests. Whereas the conventional BOD test takes 5 days, biosensors with immobilized yeast [19] or microbial arrays [20] can measure the BOD within minutes. Other areas of interest of environmental monitoring are as follows:

- Contaminants in drinking, surface, and ground water (e.g., small organic molecules)
- Detect toxic substances in seawater (e.g., heavy metals and organic pollutants)
- Monitoring volatile pollutants in industrial areas (phenols, hydrocarbons, NO_x, SO_x)
- Detect formaldehyde release (e.g., from furniture) in indoor air
- Assess soil pollution in agricultural areas (herbicides, pesticides) (Fig. 1.9)

1.5.3 Process Monitoring and Process Control

Another area of use for biosensors is monitoring and control processes in a variety of industries:

- Screening of pharmaceutical molecules (presence, extent and mechanism of activity) and presence of contaminants (inhibitors, adverse or side effects).
- Aquaculture plants monitoring (pH, oxygen, nutrients).
- Monitoring and control of fermentation processes (amino acids, aspartic acid, urea).
- Drinking water monitoring in reservoirs and pipeline networks (pH, O_2, CO_2, pollution).
- Monitoring of wastewater from industrial plants (ammonia, Cd, Cl, Zn, etc.)

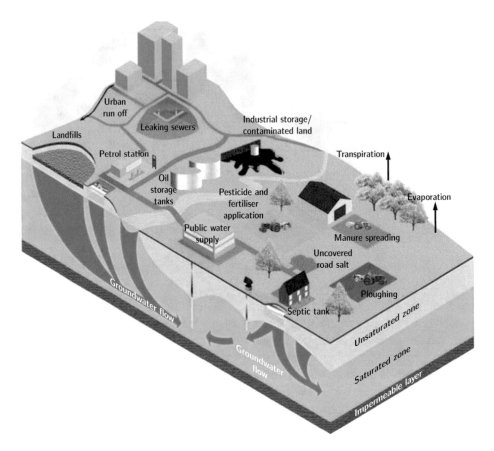

Fig. 1.9 Groundwater and different pollution paths (Source: UK Groundwater Forum, Public Domain, adapted)

1.5.4 Food Quality Control

Furthermore, biosensors can be adopted for controlling food and beverage quality [21]:

- Detection of microbiological contamination (foodborne pathogens: salmonella, *Escherichia coli*, campylobacter bacteria, listeria, vibrio cholerae, etc.).
- Detection of antibodies and mycotoxins: Naturally occurring toxins, such as mycotoxins, marine biotoxins, cyanogenic glycosides, and toxins occurring in poisonous mushrooms, periodically cause severe intoxications. Mycotoxins (e.g., aflatoxin, ochratoxin A) are found at measurable levels in many staple foods. Aflatoxin is a carcinogen.
- Detection of unconventional agents, like the ones causing the bovine spongiform encephalopathy (BSE, mad cow disease) in bovine products.

- Detection of persistent organic pollutants (POPs), like dioxins and polychlorinated biphenyls (PCBs), which accumulate in the human body and the environment.
- Degradation of adenosine triphosphate (freshness of fish, poultry, and meat).
- Detection of glucose, sucrose, methanol, ethanol (food fermentation during storage).
- Detection of citric acid, malic acid, and ascorbic acid (preservatives, additives, fruit freshness).

Question 6

Name four application areas for biosensors. How are biosensors used there?

1.6   Future Developments

The figures described above are continuously developing further; here we present some examples of what is expected to become available in the coming future.

1.6.1   Healthcare

- Development of implantable biosensors that track blood glucose levels and deliver insulin as necessary.
- Manually and wirelessly controlled biosensors that detect and treat an acute condition, being able to sense such condition and respond automatically without user intervention (also known as "in-the-flesh" physicians).
- Ultimate goal will be an implanted "lab-on-a-chip" that immediately detects illnesses, from common cold to cancer.
- Ubiquitous monitoring/control for chronically ill patient via internet.
- Wearable sensors for monitoring parameters under stress conditions and exercise [22].

1.6.2   Environmental Control and Biowarfare

- One of the most important application potentials is in industry, where biosensors can monitor air and water quality, but also emissions from chemical plants.
- Systems that detect biotoxins in just a few minutes could be deployed in sensitive public areas like subways, airports, shopping malls, and cultural sites.
- Networked wearable and smart sensors will monitor vital parameters, attention, and stress conditions of soldiers [23]. By integrating liquid-based and gas-based analysis into wearable devices with a drug reservoir and micro-controlled injectors, this type of biosensor could be incorporated into military uniforms not only for releasing alerts in

dangerous environments, but also for immediate delivery of life-sustaining drugs in case of a combat injury.

1.6.3   Food Quality Control

- Detection of glucose, sucrose, methanol, and ethanol (food fermentation during storage).
- Currently, testers take random samples from food production chains and use biosensors to detect E. coli and salmonella. Within a few years, biosensors could be used to test every single product over the whole pathway, form the production plant to the end user.

Answer 1

A biosensor is an analytical device, which converts a biological response into an electrical signal.

Answer 2

Advantages include smaller sample sizes; sensitive, selective, and specific; easy to use; faster and cheaper than traditional methods.

Applications include healthcare, pharmaceutical drug analysis, control of industrial processes, food quality control, environmental monitoring, etc.

Answer 3

The biological component recognizes a specific analyte, that is, it reacts or interacts with the specific substance to be detected.

Bioreceptors can be classified in two groups: catalytic (e.g., enzymes) and noncatalytic (e.g., antibodies).

Answer 4

In this introduction, we mentioned the following methods: amperometric, potentiometric, pH-metric, calorimetric, thermal, optical and gravimetric. The details of each method will be analyzed in the following chapters.

Answer 5

Fostering factors examples are the need of quick and cheap medical diagnosis, the rising need for environmental monitor and safe food. Also, the availability of synergistic technologies and multidisciplinary teams can facilitate the development. Last but not least, legislative activities may deliver important stimuli.

Typical hindering factors are limited reliability on the long term, limited availability of funding, regulatory legislations, and competing technologies.

Answer 6

The four most important field of application are healthcare, environmental monitoring, control and monitoring of industrial processes, food quality control.

In many areas, we observe two different approaches, namely consumer and professional products. Consumer products are typically low tech, low price, and do not require any training for proper use. Professional products include high-tech components, can be expensive, and need well-trained professional users.

Take-Home Messages
- Biosensors represent highly competitive analytical devices in comparison with the "traditional" methods.
- The fundamental components are the *bioreceptor* (biological detector) and the *physical transducer*, connected and protected by the *interfacial component*.
- The development of biosensors requires a multidisciplinary approach spanning from biology to chemistry, from physics to material science, from microelectronics to computer science, etc.
- For a commercial success, several parameters need optimization. These can be classified in six families, which behave like forces in a vector diagram, indicating the need of a good balance among them.
- Presently, especially in the field of healthcare, we find on the market two distinct classes of biosensors, namely low-tech consumer products tailored for low cost and ease of use, even at home, on the one side, and hi-tech professional products for use by trained personnel for clinical point-of-care testing on the other side.
- Most common application scenarios for biosensors are found in healthcare, industrial processes, environmental monitoring and food quality control. Other significant application fields are military and civil defense and laboratory research.

References

1. Chaplin MF, Bucke C. Enzyme technology. Cambridge: Cambridge University Press; 1990.
2. Leland C, Clark J, Wolf R, Granger D, Taylor Z. Continuous recording of blood oxygen tensions by polarography. J Appl Physiol. 1953;6(3):189–93.

3. Clark LC Jr, Lyons C. Electrode systems for continuous monitoring in cardiovascular surgery. Ann N Y Acad Sci. 1962;102(1):29–45.

4. Guilbault GG, Montalvo JG. Urea-specific enzyme electrode. J Am Chem Soc. 1969;91(8):2164–5.

5. Mosbach K, Danielsson B. An enzyme thermistor. Biochim Biophys. 1974;364(1):140–5.

6. Diviés C. Remarks on ethanol oxidation by an "Acetobacter xylinum" microbial electrode (author's transl). Ann Microbiol (Paris). 1975;126(2):175–86.

7. Lübbers DW, Opitz N. The pCO_2-/pO_2-optode: a new probe for measurement of pCO_2 or pO in fluids and gases (authors transl). Z Naturforsch C Biosci. 1975;30(4):532–3.

8. Pfeiffer EF, Beischer W, Thum C, Clemens AH. The artificial endocrine pancreas in clinical medicine and in research. Journees annuelles de diabetologie de l'Hotel-Dieu. 1976;1976:279–96.

9. Shichiri M, Yamasaki Y, Kawamori R, Hakui N, Abe H. Wearable artificial endocrine pancreas with needle-type glucose sensor. Lancet. 1982;320(8308):1129–31.

10. Liedberg B, Nylander C, Lunström I. Surface plasmon resonance for gas detection and biosensing. Sensors Actuators. 1983;4:299–304.

11. Kisner HJ. I-STAT point-of-care blood analyzing system. Clin Lab Manage Rev. 1993;7(5):454–7.

12. WHO. Diabetes: key facts. https://www.who.int/news-room/fact-sheets/detail/diabetes

13. Hoffmeier K. Pregnancy test result. https://commons.wikimedia.org/wiki/File:Pregnancy_test_result.jpg

14. Schwemmer F, Blanchet CE, Spilotros A, Kosse D, Zehnle S, Mertens HDT, et al. LabDisk for SAXS: a centrifugal microfluidic sample preparation platform for small-angle X-ray scattering. Lab Chip. 2016;16(7):1161–70.

15. The creative commons copyright licenses. https://creativecommons.org/licenses/.

16. Abbot. i-STAT family of products. https://www.pointofcare.abbott/int/en/offerings/istat

17. Healthineers S. epoc blood analysis system. https://www.siemens-healthineers.com/blood-gas/blood-gas-systems/epoc-blood-analysis-system

18. Schneider J, Dudziak R, Westphal K, Vettermann J. Der i-STAT AnalyzerEin neues, tragbares Gerät zur Bedsidebestimmung des Hämatokrits, der Blutgase und Elektrolyte. Anaesthesist. 1997;46(8):704–14.

19. Yang Z, Suzuki H, Sasaki S, Karube I. Disposable sensor for biochemical oxygen demand. Appl Microbiol Biotechnol. 1996;46(1):10–4.

20. Pitman K, Raud M, Kikas T. Biochemical oxygen demand sensor arrays. Agron Res. 2015;13:382–95.

21. Elmi M. Food safety. East Mediterr Health J. 2008;14:S143–S9.

22. Seshadri DR, Li RT, Voos JE, Rowbottom JR, Alfes CM, Zorman CA, et al. Wearable sensors for monitoring the physiological and biochemical profile of the athlete. Digital Med. 2019;2(1):72.

23. Friedl KE. Military applications of soldier physiological monitoring. J Sci Med Sport. 2018;21(11):1147–53.

Bioreceptors

2

Contents

© The Author(s), under exclusive license to Springer Nature Switzerland AG 2021
A. Pasquarelli, *Biosensors and Biochips*, Learning Materials in Biosciences,
https://doi.org/10.1007/978-3-030-76469-2_2

Keywords

Catalytic receptors · Affinity receptors · Chemical bonds · Protein structure · Steric and polar template

What You Will Learn in This Chapter
In this chapter, you will get to know the various bioreceptors. These act as the sensing component of the biosensors. You will learn about the properties of these receptors and their areas of application. We will further cover the different chemical bonds that are relevant in biochemistry. You will be able to make statements about the functionality of various types of receptors as well as their advantages and disadvantages. Furthermore, you will be able to explain the structures of a number of important biological molecular forms.

2.1   Classification of Bioreceptors

Bioreceptors are the key elements of biosensors and act as their sensing parts.
   Various biological or organic components can be used as bioreceptors:

- Enzymes: Recognition and catalysis.
- Antibodies: Recognition and binding of antigens.
- Synthetic molecularly imprinted polymers: Recognize and bind analytes.
- Cell membrane receptor: Recognition and triggering of a specific activity.
- DNA: Recognition and binding of nucleic acid strands and other entities.
- Aptamers: Synthetic biomolecules based on peptide or nucleic acid chains, work similarly to antibodies, recognize and bind specific analytes.
- Whole cells (microbes, plant or animal cells): Many of them used as natural catalytic environments, while others for studying physiologic and pathologic processes.
- Tissues or organs: Complex functional response, for example, olfactory tissue or taste buds converts chemical stimuli into action potentials.
   We now give a quick introductory look at different bioreceptors and their functionality.

2.1.1   Protein-Based Receptors

This is the most common and oldest class of bioreceptors. Here we find the three categories of enzymes, antibodies, and transmembrane proteins like gated ion channels and porins. All these molecules are natural, that is, synthetized in living organisms spanning from the simplest unicellular bacteria up to highly complex multicellular creatures like animals. Besides, it is possible to stimulate the production of specific protein receptors like antibodies, carried out by B-lymphocytes in animals, by evoking an immune response after injecting antigens [1]. Furthermore, by genetic engineering of suitable cells it is possible to obtain the synthesis of the desired proteins out of cells that would not produce such molecules in their wild-type status or also modify a protein already available in wild-type cells for fitting in a better way the targeted applications [2, 3].

Figure 2.1 depicts schematically how the three bioreceptors categories mentioned above work.

When a suitable analyte molecule binds to the receptor of a transmembrane protein like a chemo-sensitive ion channel, the channel is activated. Such channels are highly selective and allow only a specific type of ions to pass through (e.g., Ca^{2+} Ions).

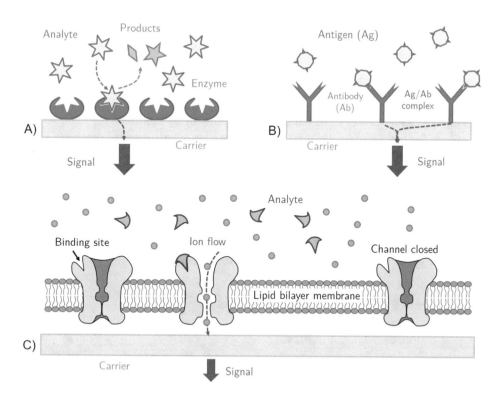

Fig. 2.1 Examples of protein-based bioreceptors: (**a**) Enzymes catalyze reactions turning a substrate, that is, the analyte to detect, into a product. (**b**) Antibody receptors bind by affinity the corresponding antigens. (**c**) Membrane receptors trigger the activity of functional transmembrane proteins, for example, ion channels. All the activities mentioned above are then sensed by an appropriate physical transducer, which delivers an output signal in a convenient format for readout and display

2.1.2 Synthetic Receptors

The gold standard for capturing receptors are the antibodies. However, they are not naturally available for every kind of analyte; therefore, in such cases, they must be expressly produced, but this approach is quite expensive and requires a considerable amount of skill. Moreover, like most functional proteins, antibodies remain stable and fully functional in a limited range of parameters like pH and temperature. Due to these reasons, there is a significant interest in developing alternatives [4]. Biomimetic chemistry deals with chemical processes that imitate biochemical reactions and is able to address the needs mentioned above. In particular, a method called *Molecular Imprinting Technique* (MIT) [5–7] is particularly convenient for the fabrication of synthetic affinity receptors and can be described as a way to create artificial "locks" for "molecular keys" (Fig. 2.2). MIT comprises different technological processes; receptors can be fabricated with a variety of methods including bulk imprinting, surface imprinting (soft lithography, template immobilization, grafting, emulsion polymerization), and epitope imprinting [6]. In most cases the receptors are fabricated with synthetic polymers, but also biomolecules like nucleic acid elements and amino acids can be conveniently adopted [8, 9].

2.1.3 DNA Receptors

The primary application of DNA receptors is capturing by hybridization the DNA strands present in the sample to be analyzed. For this purpose, single-stranded capturing DNA molecules are immobilized and aligned as arrays. Different well-known DNA sequences are attached to the spots on the DNA array, that is, on one spot there are many identical strands, on another spot are all identical strands with a different sequence and so on. Figure 2.3 shows this procedure.

As seen above, different DNA strands present in the analyte can match and hybridize with the functionalization present on different sensing spots. This method is very important in many fields of biology and medical diagnostics, ranging from species identification to

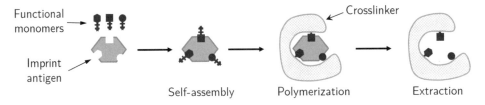

Functional monomers

Imprint antigen

Self-assembly   Polymerization   Extraction

Crosslinker

Fig. 2.2 Illustration showing the fabrication steps of an MIT process based on the Template Immobilization approach, for building an artificial affinity receptor around the target analyte, for example an antigen. Compared to natural antibodies, these receptors can be produced at low cost, are robust and durable, and have good recognition properties. However, the binding affinity is often lower compared to the one of natural molecules

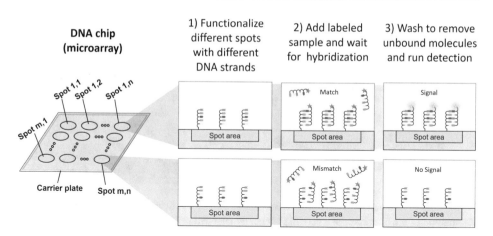

Fig. 2.3 Detection of various DNA strands at different locations on a DNA chip

forensic, from identification of genetic variations and diseases to cancer therapy, from drug screening to personalized medicine.

Furthermore, DNA receptors can be produced as synthetic receptors, as mentioned above, or also as aptamers for substituting antibodies.

2.1.4 Receptors Based on Whole Cells

This method makes use of the respiratory and metabolic functions of cells, often bacteria, to monitor analytes either as substrates or as inhibitors of these processes (Fig. 2.4). It is suitable for integration with electrochemical, optical, or calorimetric transduction techniques. We will only mention this method very briefly in this chapter.

The advantages are as follows:

- Eliminate the steps, normally required, for extraction and purification of enzymes.
- Ideal enzyme stability, thanks to the evolution-optimized biochemical environment.
- Multistep cascaded reactions are possible, which is challenging if not impossible to obtain when using a single enzyme.
- Nature provides a huge variety of microorganisms whose metabolic activities cover an enormous spectrum, thus having the potential for many biosensing applications.

At the same time, the following disadvantages need consideration:

- Multi-receptor feature may lead to lower specificity compared to single molecules.
- The stability in the long term can be challenging.
- Cells are living organisms, and therefore they need supply of oxygen and nutrients.

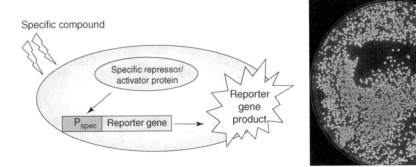

Fig. 2.4 (Left) Principle of operation of a cell as bioreceptor: An external variable triggers either the suppression or the activation of a certain genetically encoded cell activity like for example biolumi-nescence (Reprinted from [10], © 2006, with permission from Elsevier). (Right) Bacterial biolumi-nescence: Colonies of *P. mandapamensis* from the light organ of the cardinalfish Siphamia versicolor, cultivated on an agar-coated dish. (Reprinted from [11], © 2013, with permission from Springer Nature)

- The temperature range for proper function is narrow and limited.
- It is difficult to obtain a direct electrical signal and interpret it reliably.
- The response time is generally longer, because the diffusion of substrate and prod-uct molecules through the cell membrane is slower than in a simple solution.

Question 1
Can you name some advantages and disadvantages of receptors based on whole cells?

2.1.5 Tissues and Organs

Also tissues and organs can be convenient and cost-effective sources of enzyme or chemoreceptors and may offer complex and specialized recognition capabilities. These receptors provide stable and optimal activity when they are in their natural environment, where all the biomolecular pathways and cofactors are present. Immobilization over the electrode is usually done by entrapment within a semipermeable membrane. The advantages and disadvantages are similar to the case of the whole-cell elements, or even more pronounced. Suitable tissues for biosensing uses can be obtained from both plants and animals [12] (Fig. 2.5).

An example of this approach is shown in the next picture: A retina sensor, based on multisite recording of local electroretinograms (ERGs) in vitro, has been developed to

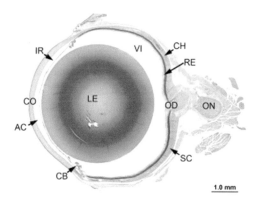

Fig. 2.5 Tissues and organs can perform complex tasks, for example, a rat's eye can be used as image sensors for studying the physiology and pathology of vision. This figure shows the histological section of a normal eye from a female F344 rat: cornea (CO), anterior chamber (AC), lens (LE), ciliary body (CB), iris (IR), choroid (CH), retina (RE), vitreous (VI), optic disk (papilla) (OD), and optic nerve (ON). (Reprinted from [13], © 2018, with permission from Elsevier)

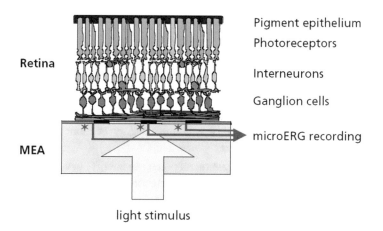

Fig. 2.6 Rat eye retina (extracted from the rat eye) is placed on a microelectrode array (MEA), making it a MEA retina sensor. In the example above, the light stimulus is projected through the translucent chip and inner retina onto the photoreceptors. The activity of ganglion cells and the local ERG are recorded by 60 microelectrodes. (Adapted from [14], © 2003, by permission from Springer Nature)

easily and effectively assess effects of pharmacological compounds on retinal activity (Fig. 2.6).

Question 2

What receptors have been introduced to you so far?

2.2   Chemical Bonds

In order to understand the detection mechanisms of bioreceptors, one should focus on their chemical and physical properties at the atomic and molecular level. Hence, it is very helpful at this point to recall the basic concepts of chemical bonds.

There are different chemical bonds or attractions, as listed below:

- Covalent bond
- Hydrogen bond (H-bridge)
- Ionic bond
- Van der Waals attraction
- Hydrophobic interaction
- Disulfide bond

The most important property of bonds is the binding energy, which is defined as the energy required to break the corresponding bond. In chemistry, this energy is measured in kilocalories per mole, recalling that: 1 kcal $= 4184$ Joule and 1 mole $= N_A$ ($\approx 6.022 \times 10^{23}$) molecules.

2.2.1   Covalent Bond

For a covalent bond, two atoms share one or more electron pairs (valence electrons) (Fig. 2.7). This is a strong kind of bond with a binding energy of about 90 kcal/mol (50–110). In comparison, the thermal energy at room temperature is only about 0.6 kcal/mol.

Covalent bonds can also store chemical potential energy that can be utilized in biological reactions. As an example of such function we can mention the phosphate anhydride bonds in adenosine triphosphate (ATP).

The angles formed between covalently bonded atoms are specific and defined. This means that biological molecules formed with covalent bonds have definite and predictable shapes, that is, 3D steric conformations.

2.2.2   Polar Covalent Bond

In the case of a polar covalent bond, for example, in water molecules (Fig. 2.8), the shared electrons are located closer to the oxygen atoms than to the respective hydrogen nucleus, because of their different electronegativity (attraction toward electrons).

Remember the electronegativity of the most common atoms in biomolecules:

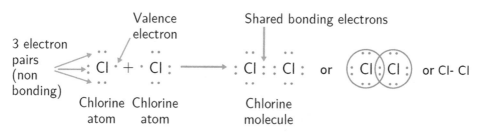

Fig. 2.7   Covalent bond between chlorine atoms

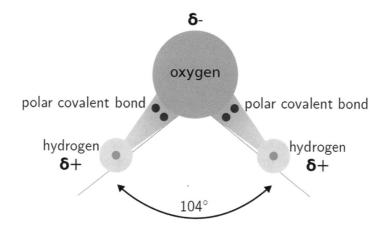

Fig. 2.8   Dipole moment at a water molecule

- Oxygen 3.5
- Nitrogen 3.0
- Carbon 2.5
- Hydrogen 2.1

Due to the resulting molecular geometry, there exists a dipole moment at water molecules. Hence, water is regarded as a polar solvent. Polar substances are correspondingly *hydrophilic*.

A similar polar molecule is ammonia (NH_3). Also, functional groups and side chains present in molecules can show a polar character, like amino ($-NH_2$), imino ($=NH$), hydroxyl ($-OH$), and carboxyl ($-COOH$) groups.

2.2.3   Hydrogen Bond

The hydrogen bond (H-bridge) is an electrostatic interaction between parts of molecules with opposite charges (Fig. 2.9). This type of bond is extremely important for stabilizing

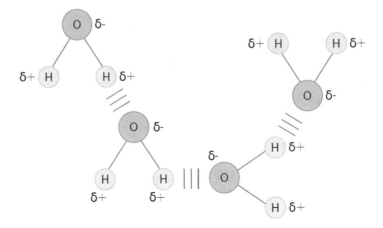

Fig. 2.9 Hydrogen bonds (H-bridges)

the structure of large macromolecules such as proteins or nucleic acids in biological systems. The bond is weak (binding energy about 5 kcal/mol) and can from or break relatively easily under normal conditions. Therefore, for stabilizing a large molecule several distributed hydrogen bonds are necessary.

2.2.4 Nonpolar Covalent Bond

In this type of bond, the common electrons are shared equally due to minimal differences in electronegativity of the atoms involved in the bond (Fig. 2.10). Hence, there is no dipole moment, and such molecules cannot enter into a polar interaction with water, thus resulting in a *hydrophobic* behavior.

The most prominent examples of nonpolar molecules are the hydrocarbons, for example, alkanes, alkenes, cycloalkanes, fats, and oils.

Fluorocarbons are another class of molecules with nonpolar character, although fluorine has an electronegativity of 4.0, which is the highest among all elements. This nonpolar behavior is the result of the steric symmetry of fluorocarbon molecules and of the extremely strong bond between fluorine and carbon atoms, resulting in short bond length and therefore in compact molecules. All these properties lead to an extreme behavior of fluorocarbons; in fact they are super-hydrophobic and also lipophobic, that is, they repel common nonpolar compounds.

Fig. 2.10 The methane
molecule is nonpolar

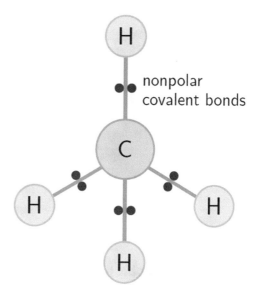

Fig. 2.11 Van der Waals bond
caused by temporary formation
of dipoles

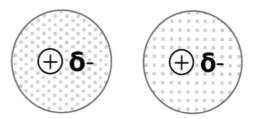

2.2.5   Van der Waals Attractions

This kind of attraction is a weak bond between nonpolar molecules. Due to the continuous movement of the shared electrons in nonpolar covalent bonds, there is a probabilistic formation of instant dipoles at any time (Fig. 2.11). These have a short lifetime and are constantly formed and lost again. The binding energy is very small (approximately 0.1–1 kcal/mol).

Van der Waals attraction is responsible for the binding forces in many gases and for the liquid and solid state of nonpolar substances.

A simple comparison between hydrogen bond and Van der Waals attraction is provided by looking at the temperature of the phase transitions of water (polar) and a simple alkane (nonpolar): Water, with a molecular mass of 18 Da, [1] melts at 0 °C and boils at 100 °C, while Octane (C_8H_{18}) with a molecular mass of 114 Da, melts at −57 °C and boils at 126 °C.

---

[1] Dalton = 1/12 of the mass of ^{12}C.

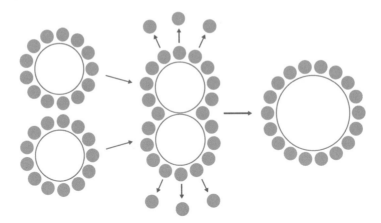

Fig. 2.12 Combination of two hydrophobic droplets is energetically favored

2.2.6   Hydrophobic Interaction

Hydrophobic molecules are insoluble in water; when mixed with water they remain in a separated phase, commonly in the form of drops, which are surrounded by the water matrix. Consider now two drops of a hydrophobic substance in water. When they come close to each other, they combine to form a greater hydrophobic region that remains excluded from the water matrix (Fig. 2.12). This combined state is energetically favorable, because less water is necessary to form the structure around the solute, and will persist. Many biological compounds are hydrophobic (e.g., cholesterol, fatty acids, and phospholipids); furthermore, large molecules with polar and nonpolar functional groups have the tendency to fold in water due to the hydrophobic interaction among the nonpolar regions.

2.2.7   Ionic Bond

The ionic bond is an attraction force between ions of opposite charges. Common compounds based on this bond are salts like the alkali halides, for example, sodium chloride, potassium bromide, etc. In biomolecules, many functional groups, like hydroxyl, amino, and carboxyl groups can form ions (Fig. 2.13). The peculiar character of ionic bonds is the different strength observed in the solid and liquid states; in fact, when a salt is in the solid state, the binding energy is similar to the one of covalent bonds, but when the same salt is mixed in a polar solvent, the binding energy drops by at least one order of magnitude.

This type of bond is important for determining the shape and structure of proteins and chromosomes. It is furthermore crucial to many reactions catalyzed by enzymes. The binding energy ranges from approximately 3–80 kcal/mol, in solution and solid state, respectively.

Fig. 2.13 Formation of ions at the carboxyl group

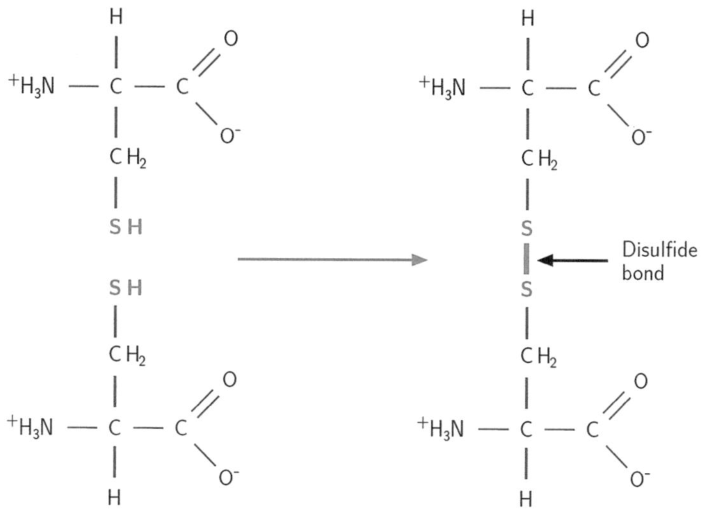

$$COOH \longrightarrow COO^- + H^+$$

negative ion   positive ion

Fig. 2.14 The disulfide bond formation between two cysteine amino acids

2.2.8 Disulfide Bond

The disulfide bond is a single covalent bond between two sulfur atoms. In biomolecules, this is the case when two cysteine amino acids in a peptide chain or protein bind together (Fig. 2.14). This type of bond is very important in determining the tertiary and quaternary structure of proteins. Furthermore, the disulfide bond plays a very prominent role in the structure of antibodies as well. The binding energy is about 60 kcal/mol.

2.2.9 Binding Energies in Water

The binding energy in an aqueous solution can differ greatly from the binding energy in a vacuum. Some values are given in the Table 2.1.

Question 3
What chemical bonds exist? How strong is their binding energy?

Table 2.1 Comparison of binding energies in aqueous solution and in vacuum

	Bond length [nm]	Binding energies in vacuum [kcal/mole]	Binding energies in water [kcal/mole]
Covalent bond	0.15	90	90
Ionic bond	0.25	80	3
Hydrogen bond	0.30	4	1
Van Der Waals bond	0.35	0.1–1	0.1

2.3   Standard Amino Acids

The main components for the synthesis of human proteins and peptides are 20 different alpha amino acids, that is, amino acids in which both the amino and the carboxyl groups are attached at the same (first, alpha) carbon atom. These 20 proteinogenic amino acids can be divided into five groups depending on their physical properties, in particular the polar and ionic behavior of the side chain (R-group), as displayed in Fig. 2.15.

2.4   Peptides, Polypeptides, and Proteins

Amino acids can form long chains by joining with peptide bonds. Here the amino group of one amino acid binds to the carboxyl group of the next amino acid, while releasing one water molecule (Fig. 2.16).

The binding energy of peptide bonds is relatively high, and amounts to about 100 kcal/mol. These bonds can even survive boiling water, but can be broken under prolonged action of acids and bases or by means of proteolytic enzymes (*proteases*) like trypsin, papain, and pepsin, among others.

The length of peptide chains ranges from just a few to several thousands of amino acids, with the molecular mass ranging from 5 to 1000 kDa. Depending on the chain length, molecules can be classified as follows (Table 2.2 and Fig. 2.17):

2.5   Proteins

This is one of the most important classes of biomolecules. Proteins are large molecules performing a vast variety of functions in living organisms. Proteins include several subclasses like enzymes, motor proteins (e.g., myosin, kinesin), ion channels and porins, transporters, antibodies, structural proteins (e.g., collagen, elastin), storage proteins (e.g., ferritin), and others. They all have in common not only the peptide bonds between amino

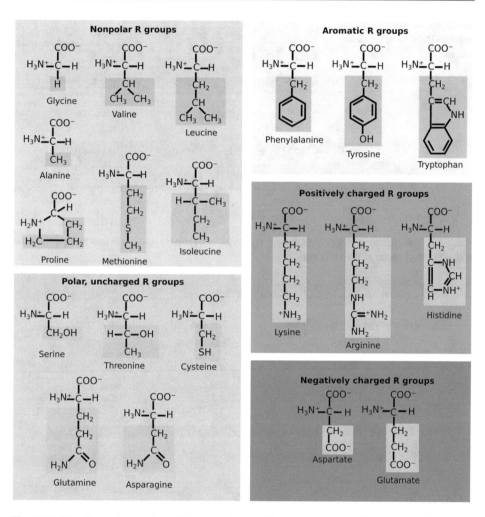

Fig. 2.15 The 20 standard amino acids that make up all human proteins. Ionization of the functional groups is explicitly indicated; however, this is dependent on the pH of the environment

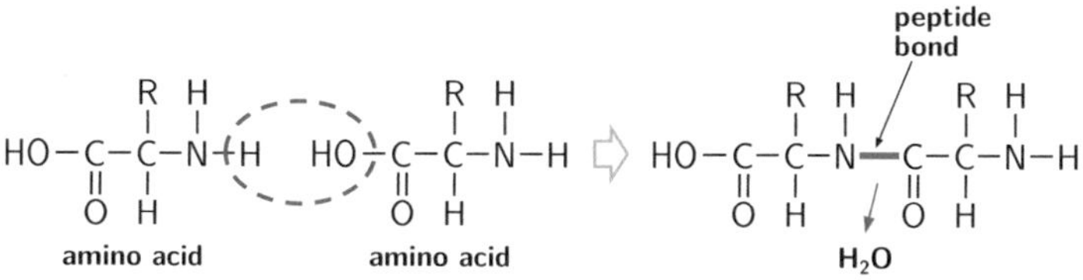

Fig. 2.16 Formation of a peptide bond between two amino acids

Table 2.2 Classification of peptides depending on chain length	Molecules classification	Number of amino acids
	Peptides (oligopeptides)	<20
	Polypeptides	<60
	Proteins	>60

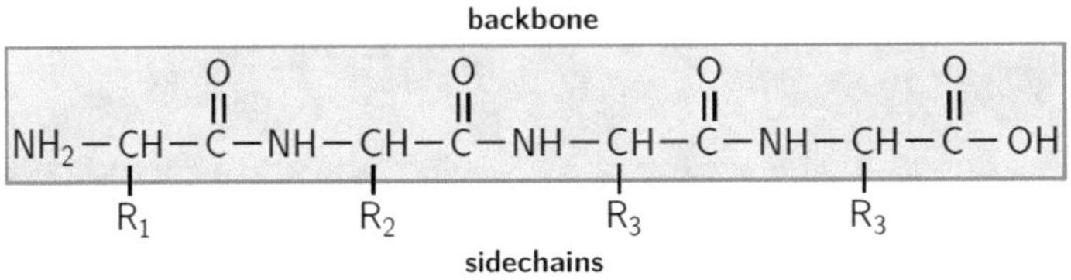

Fig. 2.17 The backbone of a polypeptide. All peptides and proteins have an amino group at one end and a carboxyl group at the other end

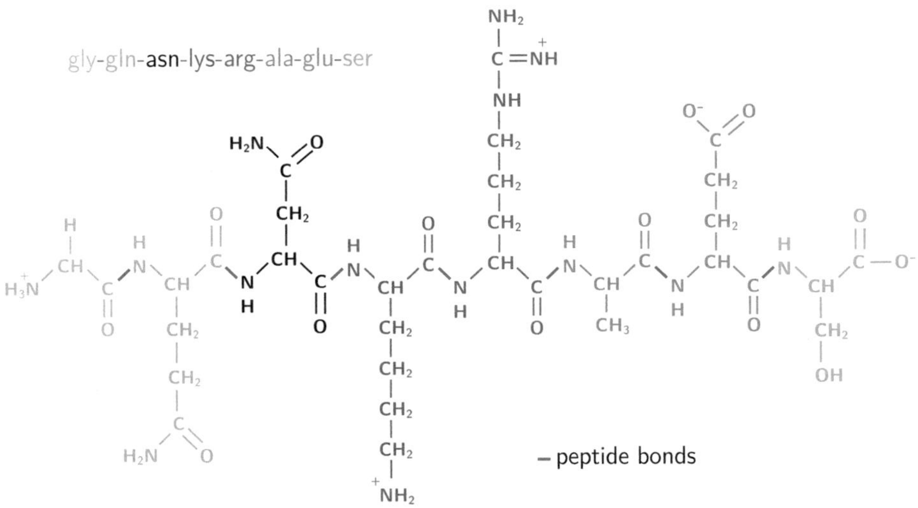

Fig. 2.18 Primary structure via peptide bonds

acids, but also the complex procedure of their structure formation, which can be classified in up to four levels. In the following, we analyze their properties.

2.5.1   Primary Structure

The primary structure of proteins is defined as the linear sequence of amino acids, linked by covalent peptide bonds. In addition, cross-links via disulfide bonds may exist. The Figures 2.18 and 2.19 show graphically these concepts.

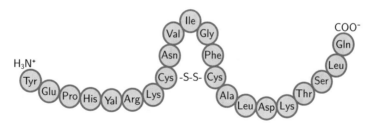

Fig. 2.19 Disulfide bond in the primary structure

2.5.2 Secondary Structure

This structure is formed due to hydrogen bonds. There are α-helix and β-sheet (Fig. 2.20). In α-helix, there is typically an H-bridge every fourth amino acid. Residues are present on the outer surface of the helix. In the β-sheet, hydrogen bonds connect parallel chains and form a folded sheet structure. This zigzag shape results from the rigid angles of the covalent bonds along the polypeptide backbone chain. For complete protein molecules, typically both α-helix and β-sheet domains are present, even multiple times, and are connected by means of simpler polypeptide chains.

Question 4

How do α-helix and β-sheet structures form in proteins?

2.5.3 Tertiary Structure

A chain with the two structures mentioned above can additionally be folded in the three-dimensional space by the interactions and bonds among the side chains of the amino acids (Fig. 2.21). This folding, driven by H-bonds, hydrophobic interactions, disulfide bonds, and ionic bonds, is in most cases a spontaneous process, which brings the molecule to a state of lower Gibbs free energy, with respect to the unfolded conformation. However, this is a dynamic condition that is influenced by environmental parameters, like pH, temperature, and ionic strength.

2.5.4 Quaternary Structure

Some very large proteins are constituted by a noncovalent "polymerization" of equal subunits (monomers) after folding in the tertiary structure. Examples of proteins with a quaternary structure are hemoglobin and transmembrane proteins like ion channels, porins, and some bacterial toxins. The figure below shows as an example Cytolysin A, a pore-

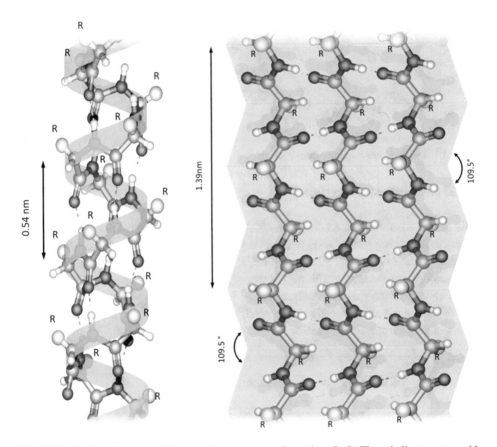

Fig. 2.20 The key elements of the secondary structure of proteins: (Left) The α-helix-structure with one turn every four amino acids. (Right) β-sheet-structure. Element atoms have the following colors: carbon gray, hydrogen white, nitrogen blue, oxygen red. The amino acid side chains are collapsed as residues (R, yellow). Hydrogen bonds, having the function of stabilizing these structures, are shown with dashed lines

forming toxin secreted by the bacterium *Escherichia coli,* in form of monomers (Colicin A) that bind to susceptible membranes forming an *Oligomer* pore (Fig. 2.22).

An effect related to the quaternary structure is the functional interaction among subunits, called *site of allosterism*. These interactions can lead for instance to the activation or inhibition of an enzyme, depending on the components involved. Using an activator "A," a protein structure with high S-affinity ("S" stands for substrate molecule, see Chap. 3) can be induced, whereas an inhibitor "I" reduces the S-affinity of the protein. This concept is illustrated in Fig. 2.23.

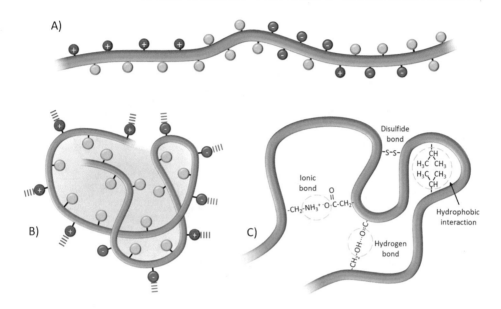

Fig. 2.21 (**a**) The effect of hydrophobic interactions on a protein chain with polar (red and blue) and nonpolar (green) side chains leads to (**b**) a hydrophobic core and a hydrophilic outer shell (**c**). The tertiary protein structure is dynamically stabilized by hydrogen bonds, disulfide bridges, ionic bonds, and hydrophobic interactions among the functional groups of the sidechains

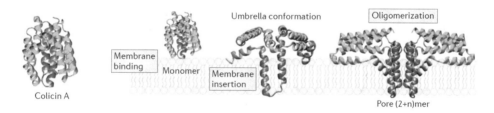

Fig. 2.22 Pore-forming toxins are proteins with a quaternary structure. (Adapted from [15] © 2008, by permission from Springer Nature)

2.6   Polarity and Ionization of Amino Acids and Proteins

The charge patterns of proteins are relatively complex and depend on environmental factors. Each carboxyl and amino group of an amino acid accepts a proton (H^+) when the pH level increases and releases it when the pH-level is decreasing, respectively (Fig. 2.24).

The acid dissociation constant K_a is the quotient of the equilibrium concentrations (in mol/L) of dissociated versus non-dissociated species, denoted by $[A^-]$, $[H^+]$ and $[HA]$, respectively.

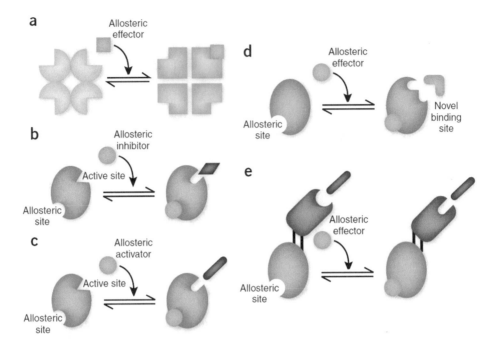

Fig. 2.23 Allosteric behaviors. (**a**) Cooperativity: A symmetric, multimeric protein can exist in one of two different states, that is, the active and inactive conformations. (**b**) A monomeric, allosterically inhibited protein. The binding of an inhibitor leads to decreased affinity or catalytic efficiency. (**c**) The binding of an activator leads to increased affinity or activity. (**d**) The binding of an effector might generate a new binding site to a protein. (**e**) Fusion of an enzyme (maroon) to a protein under allosteric control. (Reprinted from [16], © 2008, by permission from Springer Nature)

Fig. 2.24 pH-dependent association and dissociation of protons in amino acids at the amino group (left) and the carboxyl group (right)

$$pK_a = -\log K_a \rightarrow pK_a = -\log \frac{[A^-][H^+]}{[HA]}$$

pK_a and pH are related by the following relation:

$$pK_a = pH - \log \frac{[A^-]}{[HA]} \quad \text{where} : pH = -\log [H^+]$$

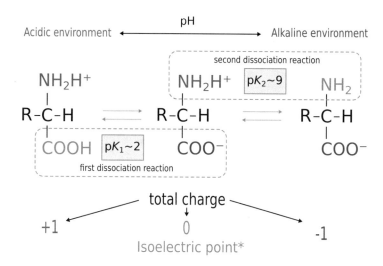

Fig. 2.25 Successive changes in the charge of an amino acid with increasing and decreasing pH-level, respectively

pK_a therefore equals pH at 50% dissociation. If the charge of amino acids depends on the pH level, several pK_a can occur.

This dynamic charge state is characterized by another parameter, namely the isoelectric point (pI), which is the pH value at which the total charge of the molecule equals 0. With the numbers given in the example shown in Fig. 2.25, it results: $pI = 5.5$.

$$pI = \frac{pK_1 + pK_2}{2}$$

Question 5

Can you explain the acid dissociation constant? How is it calculated?

A molecule that contains both negative and positive groups is called *ampholyte*. The assessment of all pK_a values and the isoelectric point pI is important for proper characterization of the amino acids. This is obtained by performing a *titration curve*, which is done by adding progressively a strong base like NaOH, while the amino acid under test is diluted in a strong nonoxidizing acid like HCl, and at the same time recording the evolution of the pH value.

As shown in Fig. 2.26, a typical titration curve of an amino acid presents inflection points, whereas the nearly horizontal ones (minimum slope) correspond to the pK_a values around which a buffering effect, that is a stabilization of the pH, is observed. The almost

Fig. 2.26 Titration curve of an amino acid. The stars indicate the pK_a values. Around the respective pK_a value, the pH remains fairly constant over a certain variation range of the base concentration. This is a buffering effect. The isoelectric point is observed at an equidistant position between the two pK_a values

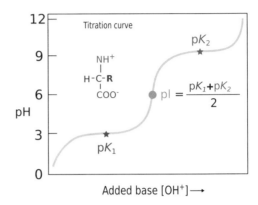

vertical inflection point (maximum slope) corresponds to the isoelectric point of that amino acid.

Some amino acids possess three pK_a values. The Table 2.3 reports the pK_a values of 20 proteinogenic amino acids.

Aspartic acid (Asp) for example possesses three pK_a values. In this case the isoelectric point pI is the average of the two K_a values, which flank the zero net-charged form (thus pK_1 and pK_2) (Fig. 2.27).

$$pI = \frac{2.1 + 3.9}{2} = 3.0$$

From the above-mentioned properties, it follows that the proteins have a very complex structure, both steric (three-dimensional object in space) and polar (charge pattern). These two characteristics are the key for understanding the selectivity and specificity of the bioreceptors, as shown in Fig. 2.28. At the same time, both the steric and the polar properties of the proteins are dependent on temperature (influence on noncovalent bonds) and pH (ionization state), and therefore their functionality is given only in a certain range of these two variables.

Question 6

How is defined the isoelectric point (pI)?

Answer 1

Advantages:

Table 2.3 List of the proteinogenic amino acids with their letter codes and respective pK_a values

Amino acid	Symbol	-COOH	-NH2	-R
Gly	G	2.34	9.60	
Ala	A	2.34	9.69	
Val	V	2.32	9.62	
Leu	L	2.36	9.68	
Ile	I	2.36	9.68	
Ser	S	2.21	9.15	
Thr	T	2.63	10.4	
Met	M	2.28	9.21	
Phe	F	1.83	9.13	
Trp	W	2.38	9.39	
Asn	N	2.02	8.80	
Gln	Q	2.17	9.13	
Pro	P	1.99	10.6	
Asp	D	2.09	9.82	3.86
Glu	E	2.19	9.67	4.25
His	H	1.82	9.17	6.00
Cys	C	1.71	10.8	8.33
Tyr	Y	2.20	9.11	10.07
Lys	K	2.18	8.95	10.53
Arg	R	2.17	9.04	12.48

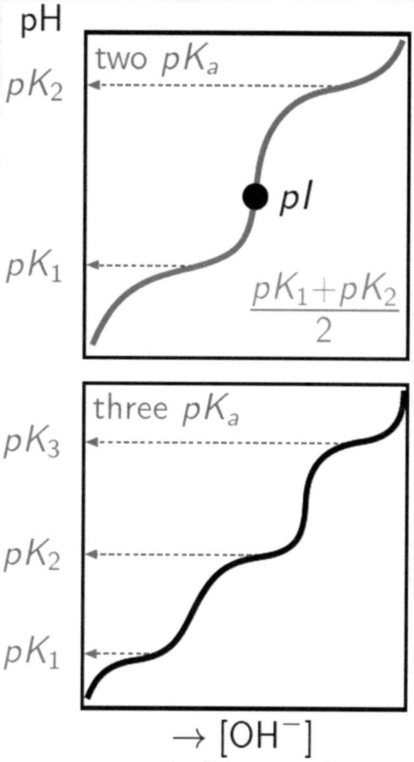

pH

two pK_a

pK_2

pI

pK_1

$\dfrac{pK_1 + pK_2}{2}$

three pK_a

pK_3

pK_2

pK_1

$\rightarrow [OH^-]$

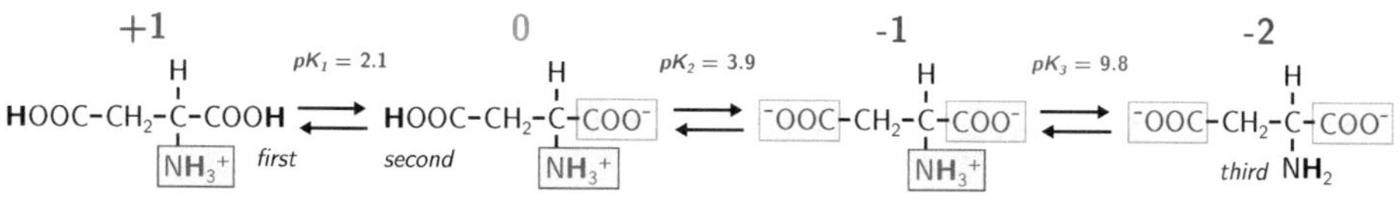

+1   0   -1   -2

$pK_1 = 2.1$ $pK_2 = 3.9$ $pK_3 = 9.8$

$$HOOC-CH_2-\overset{H}{\underset{NH_3^+}{C}}-COOH \rightleftharpoons HOOC-CH_2-\overset{H}{\underset{NH_3^+}{C}}-COO^- \rightleftharpoons {}^-OOC-CH_2-\overset{H}{\underset{NH_3^+}{C}}-COO^- \rightleftharpoons {}^-OOC-CH_2-\overset{H}{\underset{NH_2}{C}}-COO^-$$

first   second   third

Fig. 2.27 Aspartic acid (Asp) and its pK_a values

- Enzyme extraction and purification are not needed.
- Natural enzyme environment provides enhanced stability.
- Cascaded multi-enzymatic reactions easier than with isolated enzymes.

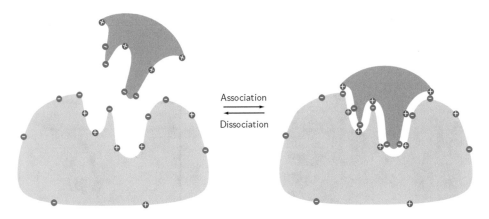

Fig. 2.28 The steric and polar structure of a protein is the key to the specific and selective noncovalent bond between receptor and analyte

- Enormous variety of microorganisms for plenty of potential applications.

  Disadvantages:

- May be less specific than single molecules.
- Long-term stability can be an issue.
- Supply of nutrients and oxygen must be provided.
- Narrower temperature range.
- Hard to obtain direct electrical signal.
- Slower response, because substrate and products must pass through the cell membrane.

Answer 2

We have seen so far:

- Protein-based receptors like enzymes, antibodies, and ion channels.
- Synthetic, biomimetic receptors fabricated via MIT.
- DNA receptors.
- Whole cells.
- Functional tissues and organs.

Answer 3

Chemical bonds can be classified as follows:

- Covalent bond, polar and nonpolar, with bonding energy of 50–110 kcal/mol.
- Hydrogen bond, with bonding energy of ~4 kcal/mol in vacuum and ~1 in water.
- Ionic bonds, with bonding energy of ~80 kcal/mol in vacuum and ~3 in water.
- Disulfide bond (covalent), with bonding energy of 60 kcal/mol.
- Van Der Waals bond, with bonding energy of ~1 kcal/mol in vacuum and ~0.1 in water.

Answer 4

α-helixes are formed when suitable domains of the peptide chain can originate a structure with a periodic hydrogen bond with every fourth amino acid along the same chain.

β-sheet structures are formed when suitable domains of the peptide chain align in parallel or antiparallel strands, which are held together and stabilized in a zigzag fashion by hydrogen bonds.

Answer 5

The acid dissociation constant K_a is the ratio of the equilibrium concentrations (in mol/L) of dissociated versus non-dissociated species.

It is calculated with the simple formula: $K_a = [A^-][H^+]/[HA]$.

Answer 6

The isoelectric point (pI) is the average of the two pK_a flanking the 0 net-charge of a molecule.

Take-Home Messages

- Most bioreceptors are protein based. In fact, enzymes, antibodies, and ion channels are proteins. Also, the sensing features present in cells, tissues, and organs belong in most cases to the three classes mentioned above.
- Additionally, DNA strands and biomimetic synthetic templates can be used as receptors in specific applications.
- We revised the chemical bonds, with special emphasis on the properties relevant for biomolecules.

(continued)

- We analyzed the chemical and physical properties of amino acids, how they bind to form peptides and proteins, and the four levels of the proteins structure.
- The steric and polar conformation of these highly complex molecules is the key for understanding how they work and why they are highly selective and specific for a given analyte.

References

1. Nossal GJV. Mechanisms of antibody production. Annu Rev Med. 1967;18(1):81–96.
2. Lee YJ, Jeong KJ. Challenges to production of antibodies in bacteria and yeast. J Biosci Bioeng. 2015;120(5):483–90.
3. Sinha R, Shukla P. Current trends in protein engineering: updates and progress. Curr Protein Pept Sci. 2019;20(5):398–407.
4. Kodadek T. Synthetic receptors with antibody-like binding affinities. Curr Opin Chem Biol. 2010;14(6):713–20.
5. Dar KK, Shao S, Tan T, Lv Y. Molecularly imprinted polymers for the selective recognition of microorganisms. Biotechnol Adv. 2020;45:107640.
6. Ertürk G, Mattiasson B. Molecular imprinting techniques used for the preparation of biosensors. Sensors. 2017;17:2.
7. Saylan Y, Yilmaz F, Özgür E, Derazshamshir A, Yavuz H, Denizli A. Molecular imprinting of macromolecules for sensor applications. Sensors. 2017;17:4.
8. Zhang Z, Liu J. Molecular imprinting with functional DNA. Small. 2019;15(26):e1805246.
9. Pavan S, Berti F. Short peptides as biosensor transducers. Anal Bioanal Chem. 2012;402 (10):3055–70.
10. Sørensen SJ, Burmølle M, Hansen LH. Making bio-sense of toxicity: new developments in whole-cell biosensors. Curr Opin Biotechnol. 2006;17(1):11–6.
11. Dunlap PV, Urbanczyk H Luminous bacteria. In: Rosenberg E, De Long EF, Lory S, Stackebrandt E, Thompson F. The prokaryotes: prokaryotic physiology and biochemistry. Berlin, Heidelberg: Springer; 2013. p. 495–528.
12. Wijesuriya DC, Rechnitz GA. Biosensors based on plant and animal tissues. Biosens Bioelectron. 1993;8(3):155–60.
13. Dunn DG, Baker JFM, Sorden SD. Eye and associated glands. In: Suttie AW, editor. Boorman's pathology of the rat. 2nd ed. Boston: Academic Press; 2018. p. 251–78.
14. Stett A, Egert U, Guenther E, Hofmann F, Meyer T, Nisch W, et al. Biological application of microelectrode arrays in drug discovery and basic research. Anal Bioanal Chem. 2003;377 (3):486–95.
15. Peraro MD, van der Goot FG. Pore-forming toxins: ancient, but never really out of fashion. Nat Rev Microbiol. 2016;14(2):77–92.
16. Goodey NM, Benkovic SJ. Allosteric regulation and catalysis emerge via a common route. Nat Chem Biol. 2008;4(8):474–82.

Catalytic Biosensors

3

Contents

Keywords

Catalysis · Reaction kinetics · Inhibition · Calorimetric detection · Electrochemical detection methods

What You Will Learn in This Chapter
In this chapter, you will learn about catalytic reactions and biosensors that make use
of these reactions. Calorimetric sensors detect the release of heat, whereas electrochemical sensors detect the release of charge carriers in a biochemical reaction. You
will be able to explain the functionality and characteristics of these sensors, as well as
select the appropriate technology for each application.

Biochemical reactions can be influenced by the presence of catalysts, such as enzyme
molecules. Enzymes are able to speed up the kinetics, whereas external factors can induce
inhibition or activation. Further influential factors exist, for example the pH value and the
temperature. In practical sensing applications, enzymes can be used as bioreceptors in
calorimetric and electrochemical biosensors.

3.1   Catalysis of a Reaction

The rate of a reaction increases significantly upon reduction of the energy barrier through
catalysts (Fig. 3.1). For example, the uncatalyzed breakdown of hydrogen peroxide to
water and oxygen has an activation energy $\Delta G_U = 76$ kJ/Mol. This energy barrier is
reduced to $\Delta G_C = 30$ kJ/Mol in the presence of the enzyme catalase. The reaction rate
hereby increases by a factor of 10^8 due to the exponential dependencies of thermodynamic
processes.

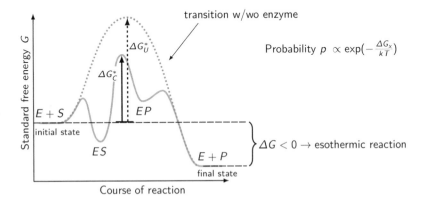

Fig. 3.1 Energy diagram of an exothermic reaction, with and without enzymatic catalysis. The
course of the catalyzed reaction consists of four states: free enzyme and substrate $E + S$, formation of
the enzyme–substrate complex ES, turn the substrate into product within the complex EP, and release
of product and free enzyme $E + P$. By catalysis, the energy barrier drops from ΔG_U to ΔG_C

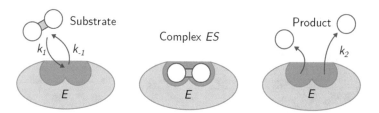

Fig. 3.2 Illustration representing an enzymatic reaction

In fact:

$$kT = 2.479 \text{ kJ/Mol (thermal energy at room temperature)}$$

the probability for the occurrence of the uncatalyzed reaction is proportional to

$$\exp\left(-\Delta G_U/kT\right) = \exp\left(-76/2.479\right) = \exp\left(-30.5\right) = 5.67 \times 10^{-14}$$

while the probability for the occurrence of the catalyzed reaction is proportional to

$$\exp\left(-\Delta G_C/kT\right) = \exp\left(-30/2.479\right) = \exp\left(-12.1\right) = 5.56 \times 10^{-6}$$

From a practical point of view, a reaction time that is measured in years without catalysis can be measured on the seconds scale if catalyzed by the enzyme.

The simplest scheme for describing a biochemical reaction catalyzed by a single enzyme E (Fig. 3.2) is the irreversible conversion of a substrate S to a product P through the reversible formation of an intermediate enzyme–substrate complex ES:

$$E + S \underset{k_{-1}}{\overset{k_1}{\rightleftarrows}} ES \overset{k_2}{\rightarrow} E + P$$

Here the velocity of reaction $X \overset{k_x}{\rightarrow} Y$ is given by:

$$V_x = \frac{d[Y]}{dt} = k_x[X]$$

k_1, k_{-1} and k_2 are the respective rate constants and typically have values of 10^5–10^8 s^{-1}, 1–10^4 s^{-1}, and 1–10^5 s^{-1}. For large values of the substrate concentration $[S]$ ($k_1 >> k_{-1}$, k_2), $[ES]$ quickly reaches the saturation condition and then remains almost constant (steady state condition).

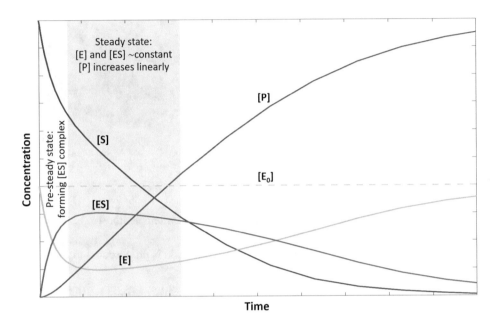

Fig. 3.3 Time course of an enzymatic reaction with respect to the reactants concentrations

The equilibrium constant K_m is called *Michaelis constant* and is given by the ratio of the total breakdown rate $(k_{-1} + k_2)$ and the formation rate (k_1) of ES:

$$K_m = \frac{k_{-1} + k_2}{k_1}$$

At the steady state, we use the following approximation:

$$[S_0] \gg [E_0] \rightarrow \frac{d[ES]}{dt} \approx 0$$

Here $[E_0]$ and $[S_0]$ are respectively the total enzyme concentration and the substrate concentration at the beginning, whereas $[E]$ and $[S]$ are the concentrations of free enzymes and substrate at any given time (Fig. 3.3). Therefore,

$$[E] = [E_0] - [ES]$$

At very high substrate concentrations, the reaction rate is approximately constant and the decrease of $[S](t)$ from its initial value $[S_0]$ is linear (Fig. 3.4). It is a zero-order reaction:

Fig. 3.4 Reaction
velocity vs. substrate
concentration

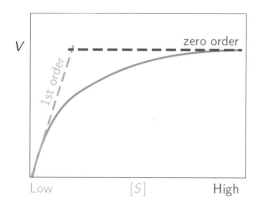

$$[S](t) = [S_0] - V_{\text{max}}t$$

At very low substrate concentrations, the reaction rate is almost linearly dependent on the concentration (first-order reaction). Hence $[S](t)$ is given by:

$$[S](t) = [S_1]e^{-\left(\frac{V_{\text{max}}t}{K_m}\right)}$$

$[S_1]$ is the substrate concentration below which the reaction kinetics is almost first order.

Question 1
Why are enzymes used in a biochemical reaction?

3.1.1   Michaelis–Menten Equation

The reaction velocity is given as the product of $[ES]$ and k_2:

$$V = \frac{d[P]}{dt} = k_2[ES]$$

$[ES]$ depends on the rate of formation (k_1) and breakdown (k_{-1}, k_2):

$$\frac{k_1([E_0] - [ES])}{[S]} = k_{-1}[ES] + k_2[ES]$$

After a few transformations, we obtain:

$$[ES] = \frac{k_1[E_0][S]}{k_1[S] + k_{-1} + k_2} = \frac{[E_0][S]}{[S] + \frac{k_{-1}+k_2}{k_1}}$$

Recalling the Michaelis constant K_m, we get:

$$[ES] = \frac{[E_0][S]}{[S] + K_m} \rightarrow V = \frac{k_2[E_0][S]}{[S] + K_m}$$

With the maximum reaction velocity $V_{max} = k_2[E_0]$, which is reached when the enzyme is fully saturated (i.e. $[ES] = [E_0]$), we obtain finally the Michaelis–Menten equation:

$$V = \frac{V_{max}[S]}{[S] + K_m}$$

When solving the Michaelis–Menten equation for $V = 0.5\ V_{max}$ we obtain:

$$\frac{V_{max}}{2} = \frac{V_{max}[S]}{[S] + K_m} \rightarrow \frac{1}{2} = \frac{[S]}{[S] + K_m}$$

and after rearranging the terms:

$$[S] + K_m = 2[S] \rightarrow K_m = [S]$$

this means, that K_m can be experimentally determined as the concentration for which we measure $V = 0.5V_{max}$. This results is represented graphically in Fig. 3.5.

A better representation of the Michaelis–Menten equation is obtained by forming the double reciprocal equation:

Fig. 3.5 Graphic representation of Michaelis–Menten equation

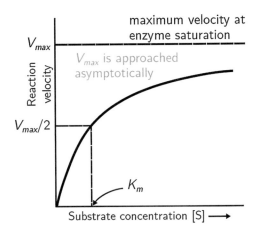

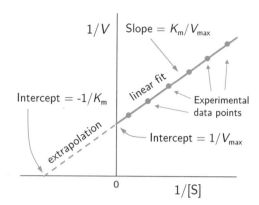

Fig. 3.6 Lineweaver–Burk plot (double reciprocal representation of Michaelis–Menten equation)

$$\frac{1}{V} = \frac{[S] + K_m}{V_{max}[S]} \rightarrow \frac{1}{V} = \frac{1}{V_{max}} + \frac{K_m}{V_{max}} \frac{1}{[S]}$$

Through that linearization (here the variables of a linear equation in the form $y = ax + b$ are $y = 1/V$ and $x = 1/S$), we get a more precise experimental determination of the reaction parameters V_{max} und K_m, which are obtained from the intercepts (Fig. 3.6). Note that in the left half plane the function is extrapolated down to the intercept (negative values of $1/[S]$ are impossible!).

3.1.2 Inhibition

Enzyme inhibition can occur in the presence of an inhibitor and this will influence the reaction kinetics. If the inhibitor competes with the substrate when docking with enzyme molecules, it reduces the reaction velocity. This is called *competitive inhibition* (Fig. 3.7 top).

A reduction in reaction velocity can also take place when the inhibitor does not compete with the substrate molecules for docking at the active center of the enzyme, but is able to dock at another different site. This leads to the formation of an *EIS* complex, which in turn lowers the concentration of *ES*. [*ES*] is however decisive for the formation of product *P*. This is called *noncompetitive inhibition* (Fig. 3.7 bottom).

3.1.3 Michaelis–Menten Equation for Competitive Inhibition

When solving the Michaelis–Menten equation, we observed that the total enzyme concentration can be expressed as

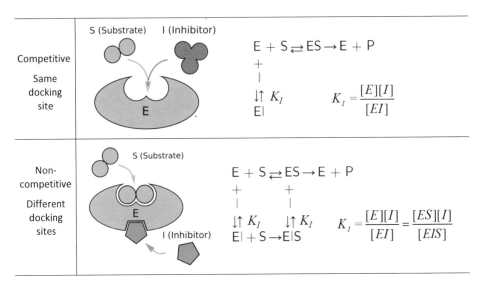

Fig. 3.7 (Top) Competitive inhibition: [*I*] binds to free [*E*] only, and competes with [*S*], reducing the availability of the enzyme for the target reaction. Increasing [*S*] overcomes the inhibition. (Bottom) Noncompetitive inhibition: [*I*] binds to free [*E*] or [*ES*] complex and partly subtracts the enzyme from the balance of the ES complex formation. Increasing [*S*] cannot overcome inhibition

$$[E_0] = [E] + [ES]$$

Now, in the presence of a competitive inhibitor we have

$$[E_0] = [E] + [ES] + [EI]$$

Solving the Michaelis–Menten equation such that this variation is taken into consideration, leads to the following equation:

$$V = \frac{V_{max}[S]}{[S] + \alpha K_m} \quad \text{where :} \quad \alpha = 1 + \frac{[I]}{K_I} \text{ and } K_I = \frac{[E][I]}{[EI]}$$

Thus, in the absence of an inhibitor, the value of α is 1 and we get the original equation. In the presence of an inhibitor, the value of α is larger. The higher the inhibitor concentration compared to K_I, the higher α.

Thus, in the presence of a competitive inhibitor, the effective $K_m{}'$ (i.e., αK_m) will be higher.

Since the substrate competes with the inhibitor, it is evident that for [*S*] much larger than $K_m{}'$ (i.e. [*S*] conveniently large), the inhibitor does not affect the measurements, therefore V_{max} does not change. However, the presence of a competitive inhibitor requires a higher

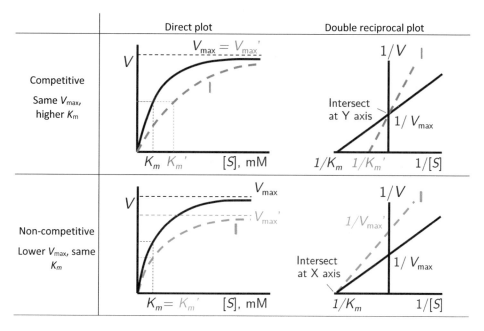

Fig. 3.8 The influence of inhibitors on the reaction kinetics

substrate concentration to achieve the same (half) maximal rate observed in the uninhibited case (Fig. 3.8 top).

3.1.4 Michaelis–Menten Equation for Noncompetitive Inhibition

In the presence of a noncompetitive inhibitor, we have:

$$[E_0] = [E] + [ES] + [EI] + [EIS]$$

Solving the Michaelis–Menten equation on this basis, we obtain:

$$V = \frac{V_{max}[S]}{\alpha[S] + \alpha K_m} = \frac{V_{max}}{\alpha} \frac{[S]}{[S] + K_m} \quad \text{where again}: \quad \alpha = 1 + \frac{[I]}{K_I} \text{ and } K_I = \frac{[E][I]}{[EI]}$$

This means that in presence of a noncompetitive inhibitor, $K_m' = K_m$, that is, it is invariant. In this case, the reaction rate scales down with α at all substrate concentrations $[S]$; this means that there is no way to recover V_{max} in the presence of a noncompetitive inhibitor (Fig. 3.8 bottom).

Question 2

What does the Michaelis–Menten equation describe?

Question 3

How does the inhibitor influence the reaction kinetics?

3.2   Enzyme Activity

The international unit for enzyme activity (IU) is defined as the amount of enzymes that catalyzes 1 μM of substrate per minute under standard conditions of temperature, optimal pH, and substrate concentration. Quite often, the activity is indicated in terms of turnover rate, that is, as the number of substrate molecules that an enzyme molecule converts into product per second. Table 3.1 lists some enzymes and their activity (NB, values are not in IU!).

The enzymatic activity is strongly influenced by factors such as pH, temperature, and mechanical stress or pressure. Deviation from optimal conditions may result reversible (i.e., by restoring the normal environment, the enzyme goes back to its original functionality), but in some cases, like extreme temperatures, the change is irreversible due to permanent denaturation (Fig. 3.9).

Some enzymes, which have a quaternary structure, consist of several identical polypeptide chains (monomers), each of which may have its own active site. Such enzymes can therefore bind more than one substrate molecule simultaneously. Activators and/or inhibitor as specific molecules that can dock on allosteric enzymes, thus stabilizing the active or inactive form respectively (Fig. 3.10).

Catalytic reactions can induce a variety of effects, as shown in Fig. 3.11. In the following, all these methods will be analyzed systematically.

Table 3.1 Maximum turnover numbers of some enzymes (Source [1])

Enzyme	Turnover rate [molecules/s]
Carbonic anhydrase	600,000
3-Ketosteroid isomerase	280,000
Acetylcholinesterase	25,000
Penicillinase	2000
Lactate dehydrogenase	1000
Chymotrypsin	100
DNA polymerase I	15
Tryptophan synthetase	2
Lysozyme	0.5

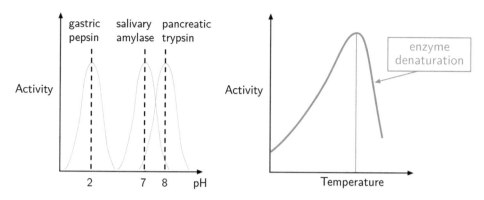

Fig. 3.9 Factors influencing the enzyme activity: (left) pH, (right) temperature

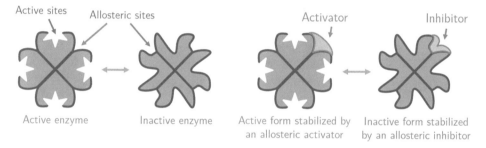

Fig. 3.10 (Left) Active and inactive forms of a polymeric enzyme. (Right) An activator or an inhibitor molecule is able to stabilize the given specific state

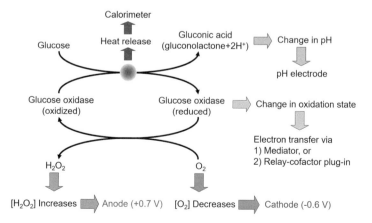

Fig. 3.11 Several options may be available for detecting a catalytic reaction

3.3   Calorimetric Biosensors

Many reactions catalytically influenced by enzymes are exothermic, that is, heat is produced during the reaction. A calorimeter can be used to measure the reaction rate and analyte concentration.

Table 3.2 shows the heat released by some of the enzymes used for biosensing

Inside a calorimeter (Fig. 3.12), the sample solution S (1) flows through the thermal insulation (2) reaching a heat exchanger (3) embedded in an isothermal block (4). Further, the solution passes by the first thermistor, which acts as reference (5) and immediately after

Table 3.2 Heat release by some enzymatic reactions (Source: [2])

Enzyme	Reactant	Heat release $-\Delta H$ [kJ/Mol]
NADH dehydrogenase	NADH	225
β-Lactamase	Penicillin G	115
Catalase	Hydrogen peroxide	100
Glucose oxidase	Glucose	80–100
Hexokinase	Glucose	75
Lactate dehydrogenase	Sodium pyruvate	62
Urease	Urea	61
Cholesterol oxidase	Cholesterol	53
Uricase	Urate	49
Trypsin	Benzoyl-L-arginine amide	29

Fig. 3.12 Scheme of a calorimetric biosensor

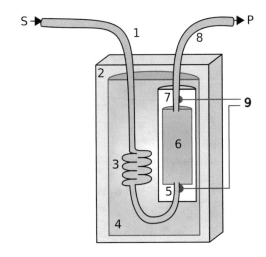

enters the bioreactor (6), which contains the enzyme. The catalytic exothermic reaction happens here and releases heat. The second thermistor (7) measures the output temperature and then the solution P containing the reaction products leaves the sensing system (8). An externally placed read-out electronics (9) measures the deviation of resistance in the thermistors, thus assessing the temperature difference and hence the amount of released heat.

NTC thermistors have in first approximation a resistive value R proportional to $e^{-\left(\frac{\alpha}{T}\right)}$, where α, a material-related parameter, is the temperature constant of the thermistor and T the absolute temperature. On this basis we can write:

$$\ln\left(\frac{R_1}{R_2}\right) = \alpha\left(\frac{1}{T_1} - \frac{1}{T_2}\right) \rightarrow \frac{R_1}{R_2} = \exp\left[\alpha\left(\frac{1}{T_1} - \frac{1}{T_2}\right)\right]$$

By approximating $e^x \approx 1 + x$ for $|x| \ll 1$ we get:

$$R_1 = R_2\left[1 + \alpha\left(\frac{1}{T_1} - \frac{1}{T_2}\right)\right]$$

Since the variation of temperature is generally small, that is, $T_1 \approx T_2$, we can assume for the thermistors $R_1 \approx R_2$, thus we can simplify the above relation as:

$$\frac{\Delta R}{R} = -\frac{\alpha}{T_1^2}\Delta T$$

A typical value for the sensitivity parameter $-\alpha/T_1^2$ is $-4\%\ °C^{-1}$.

Close matching of thermistors temperature coefficients is crucial. A variation of 1 °C in the temperature of the isothermal block, that is, a common-mode drift, may cause a measurement artifact equivalent to 0.01 °C, which is of the same order of magnitude of the system full-scale range. However, the thermal insulating case and the isothermal block effectively minimize the influence of environmental temperature changes by thermally insulating and stabilizing the reactor.

Both the sensitivity (0.1 mM) and the measuring range (0.1–10 mM) of the calorimetric biosensors are very limited for most applications. However, the sensitivity can be improved by cascading several reactions in the same pathway, thus increasing the cumulative heat output. For example, in case of glucose detection with a glucose oxidase test (GOD test), in addition to glucose oxidase, the amount of heat can be increased by adding catalase to convert the hydrogen peroxide to water and oxygen. The total amount of heat is then the sum of the quantities of heat produced by glucose oxidase and catalase. The sensitivity can thereby be more than doubled.

An extreme example of such enhancements of sensitivity is depicted in the scheme of Fig. 3.13 representing a reaction for the detection of ADP. Four enzymes are

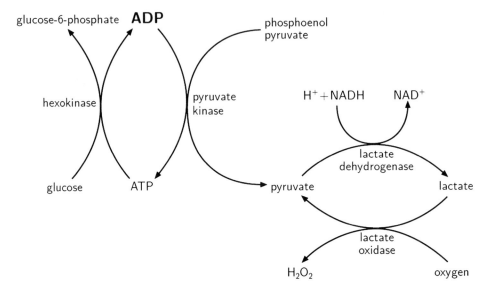

Fig. 3.13 Amplification of heat release by a factor of ~1000, by cascading four reactions. The use of kinases makes available a large energy budget by adding/removing phosphate groups to/from the substrates. This is called *phosphorylation/dephosphorylation*

co-immobilized within the reactor. With this method, the overall sensitivity is increased by three orders of magnitude [3].

Question 5

Name a few limitations of calorimetric biosensors!

How can be increased the sensitivity of these sensors?

3.4 Electrochemical Biosensors

The most important physical transduction methods for catalytic biosensors are based on electrochemistry. In the following sections, we will take a closer look at the properties and functionality of various approaches for the realization of these sensors.

3.4.1 Electrode Properties

In electrochemical biosensors, the electrodes play the role of the physical transducer. There are two main categories of electrodes suitable for biosensing applications. In the so-called *Type 0* (redox) electrodes, ion exchange with the electrolyte does not take place. In fact,

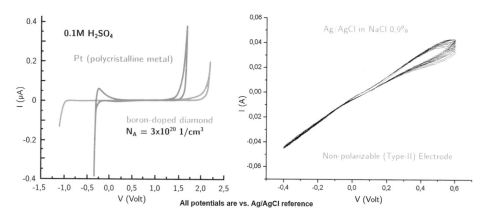

Fig. 3.14 Cyclic voltammograms (I vs. V plots). (Left) Type 0 electrodes. The peaks at the ends of the plots indicate water electrolysis, with hydrogen evolution in the cathodic range and oxygen evolution in the anodic range. At intermediate potentials, a current is detected only in the presence of reducible/oxidizable molecules, otherwise the current intensity is negligible. (Right) Type II electrode. This I–V characteristic shows a good linearity over multiple scan cycles. The current drop observed in the upper anodic range is due to diffusional limitation by collecting Cl^- ions from the electrolyte for building the AgCl layer. This limitation is not observed in the reverse direction

when a bias voltage is applied, these electrodes polarize, that is, they behave similarly to capacitors. An electrical current, in the form of direct electron exchange only occurs when an electrochemical reduction–oxidation (redox) reaction takes place at their surface. When connected to a current amplifier, such electrodes deliver precise information on the reaction rate. Although these electrodes are not subject to wear and tear, they can have reduced activity and/or a significant parasitic leakage current due to accumulation of reaction products at the surface. This phenomenon is called *fouling*. In such cases, the electrodes should be thoroughly cleaned, with a choice of mechanical, wet-chemical or dry-chemical methods.

Typical materials for the production of Type 0 electrodes are noble metals (gold and platinum) and all carbon-based materials (graphite, amorphous carbon, and boron-doped diamond) (Fig. 3.14 left).

In contrast, *Type II* electrodes are not polarizable, that is, they behave similarly to resistors. Charges are transferred here via ion exchange, between the solid electrode and the electrolyte. Type II electrodes respond fairly linearly to voltage changes (Fig. 3.14 right). They consist of a quasi-noble metal coated with one of its salts, which is hardly soluble in the electrolyte. Examples are silver coated with silver chloride (Ag/AgCl) and mercury coated with mercury chloride (Hg/HgCl). The main error source is temperature drift. Type II electrodes find use in high-impedance potentiostats since they allow the precise measurement of electrochemical potentials.

Type II electrodes undergo consumption of their salt layer; therefore, this layer needs to be regenerated at appropriate intervals depending on the cumulative ion exchange. Such regeneration can be passive, like immersing a silver electrode in sodium hypochlorite (NaOCl) for several hours, or shortly deepening it into melted silver chloride (at $T > 455\,°C$), or by actively impressing an anodic current by keeping the silver element immersed in a hydrochloric acid (HCl) or potassium chloride (KCl) solution. The latter procedure takes about 10–30 min, depending on the current density.

A last remark about electrodes: There are also *Type I* electrodes, consisting of simple metals like copper or zinc which directly exchange ions (Cu^{2+} and Zn^{2+} respectively) with the electrolyte. Since in most cases these metal ions are not present in biological milieus, this kind of electrodes are generally not appropriate for biosensing applications.

Question 6

What is the difference between Type 0 and Type II electrodes in an electrochemical biosensor?

What are their respective properties?

3.4.2 Potentiostat

Precise electrochemical measurements are usually performed using a potentiostat. A typical construction scheme of such instrument is shown in Fig. 3.15.

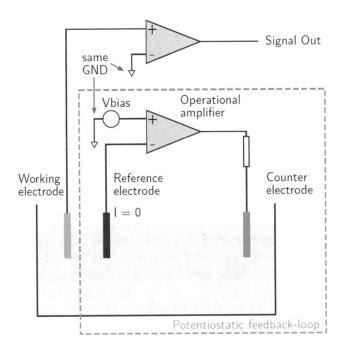

Fig. 3.15 Potentiostat with a three-electrode setup consisting of a working-, counter-, and a reference-electrode (usually indicated as WE, CE, and RE, respectively)

If the potentiostat is used for potentiometry, the upper operational amplifier is configured in voltage mode (voltage follower/amplifier, input impedance $Z_{in} \approx \infty$). On the other hand, if the potentiostat works for amperometry, the amplifier operates in current mode (transimpedance amplifier, $Z_{in} \approx 0$).

The reference electrode is a Type II electrode immersed in a cuvette filled with a nominal solution and contacted with the main electrolyte via a pinhole or a porous feed-through called **frit**. Thanks to this construction, the reference electrode provides a reliable nominal potential and is operated without current to foster an ideal behavior. The working electrode is a Type II electrode for potentiometry and a Type 0 electrode for amperometry. The counter electrode is typically made of platinum, graphite, or carbon.

3.5 Potentiometric Biosensors

Potentiometric biosensors detect the change in electrochemical potential, usually at constant zero current, generated by the biocatalytic reaction. In the simplest case, they are based on an analyte-permeable membrane, which holds immobilized enzymes wrapped around a glass membrane pH meter, while a catalyzed reaction generates or absorbs hydrogen ions, causing a pH change (Fig. 3.16). Such electrode setups have a very high

Fig. 3.16 Scheme of a simple potentiometric biosensor

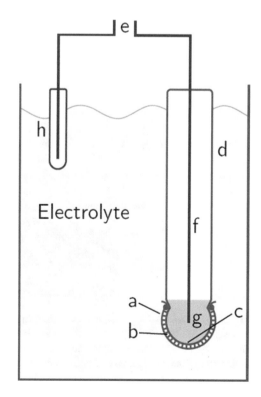

impedance and thus do not effectively allow current flow. Accordingly, they do not interfere with the reaction.

A simple potentiometric biosensor is illustrated in the figure above: A semipermeable membrane (**a**) wrapped around the cuvette tip entraps the biocatalyst (**b**) holding it close to the ion-selective thin glass membrane (**c**) of a pH sensor (**d**). The potential (**e**) is measured across the built-in Ag/AgCl electrode (**f**), which is immersed in a nominal HCl solution (**g**) and a reference electrode (**h**) immersed in the external electrolyte.

In biosensors, we can classify three types of ion-selective electrodes:

1. Glass electrodes for positive ions, that is, cations (like H^+ and NH_4^+), which are fabricated using a very thin hydrated glass membrane. The ion-selectivity is obtained by tailoring the composition of the glass. An example is the usual pH electrode.
2. pH-sensitive glass electrodes like above, but with an additional gas-permeable membrane specifically selective for gases like CO_2, H_2S, or NH_3. The gas diffusion through such membrane leads to a change in pH, which in turn is detected as electrical potential.
3. Solid-state Ion-Sensitive Field-Effect Transistor (ISFETs) fabricated by means of microelectronic technologies: The channel of such a planar device is covered with a thin, metal-based or polymer-based ion-selective membrane (Fig. 3.17). Palladium is commonly adopted for H^+, while doped polyvinylchloride works well with alkaline ions. Also, polyacrylamide or a mixture of silver sulfide and silver halide are frequently

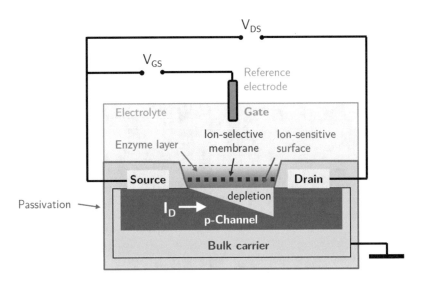

Fig. 3.17 Scheme of a potentiometric biosensors based on a solid-state Ion-Sensitive Field-Effect Transistor (ISFET). The charge of the ions resulting from the catalytic reaction and (optionally) selected by an appropriate membrane induces a depletion of charge carriers in the FET channel, which modulates the channel resistance and hence the current intensity

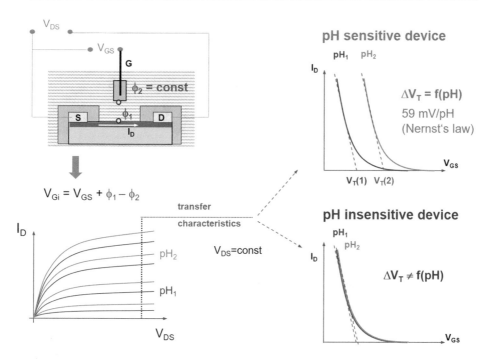

Fig. 3.18 Scheme and electrical characteristics of an ISFET at different pH values in comparison with a pH-insensitive device. The effective gate voltage V_{Gi}, which modulates the drain current, is given by the sum of the biasing potential V_{GS} applied at the solid gate, minus the electrochemical potential of the bulk electrolyte ϕ_2 plus the variation of the local electrochemical potential ϕ_1 resulting from the activity of the biocatalytic layer and the selective membrane. Ion-insensitive devices respond to variations of V_{GS} only (Courtesy: Andrej Denisenko, University of Stuttgart)

utilized. The solution + reference electrode is equivalent to the gate of a FET. It is often called *liquid gate*.

An ISFET behaves like a junction-gate field-effect transistor (JFET), with the difference that the gate potential is provided in the form of an electrochemical potential by the electrolyte (liquid gate), thus it is ion-sensitive (Fig. 3.18). This potential has two contributions:

1. The bulk electrolyte is biased at a fixed potential by means of an external potentiostat.
2. Local reactions induce deviations from the preset bias.

As mentioned above, a selective membrane can be added on top of the channel to make the device sensitive to specific ion species only. In such cases, we speak of ion-selective FET, like pH-FET (only H^+ sensitive) or K-FET (sensitive only to K^+).

The responsive behavior of an ion-selective electrode is given by the Nernst equation:

Table 3.3 Potential values for common reference electrodes

Electrode	Potential E_0 (V) at 25 °C	Temperature coefficient (V/°C) around 25 °C
Standard hydrogen electrode (SHE)	0.000	+0.00087
Ag/AgCl in 3.5 Mol/kg KCl	+0.205	−0.00073
Ag/AgCl in 1.0 Mol/kg KCl	+0.235	+0.00025

$$E = E_0 + \frac{RT}{zF} \ln [i]$$

where E is the measured potential in volt, E_0 is a characteristic constant, that is an internal potential offset, for the ion-selective/external electrode system (Table 3.3), R is the gas constant (8.314 J/mol K), T is the absolute temperature (K), z is the signed ionic charge, F is the Faraday constant (96,485 C/mol), $[i]$ is the activity ($\approx$ concentration) of the free uncomplexed ionic species.

We now look at the Nernst equation in more detail for redox reactions, from an energetic point of view. The chemical potential μ_c of a solution of redox-active molecules is the difference between the energy barriers for taking electrons from and for giving electrons to the working electrode that is setting the solution's electrochemical potential ($\mu_c = eE_c$).

$$\frac{[Ox]}{[Red]} = \frac{\exp\left(-[\text{Energy barrier for electron loss}]/kT\right)}{\exp\left(-[\text{Energy barrier for electron gain}]/kT\right)} = \exp\frac{\mu_c}{kT}$$

with

$$[Ox] + e^- \rightleftarrows [Red]$$

The ratio of oxidized to reduced molecules ([Ox]/ [Red]) is equivalent to the probability of being oxidized (giving electrons) over the probability of being reduced (taking electrons), which we can write in terms of the Boltzmann factors for these processes. Taking the natural logarithm of both sides gives

$$\mu_c = kT \cdot \ln \frac{[Ox]}{[Red]}$$

If $\mu_c \neq 0$ for $[Ox]/[Red] = 1$, we have to add an additional constant: $\mu_0 = eE_0$.

We now divide both sides by e, to convert the chemical potential into the electrode potential and remembering that $kT/e = RT/F$,[1] we obtain the Nernst equation for a reaction which exchanges z *signed* elementary charges (ions or electrons):

$$E = E_0 - \frac{RT}{zF} \ln \frac{[\text{Red}]}{[\text{Ox}]}$$

At room temperature, under transformation of the natural log to log in base 10, we obtain:

$$E = E_0 - \frac{0.0592}{z} \log \frac{[\text{Red}]}{[\text{Ox}]}$$

This gives us a theoretical sensitivity of 59.2 mV/pH in the measured potential at a temperature of 25 °C. The logarithmic dependence on the ionic concentration in the Nernst equation is responsible for both the wide analytical range and the low accuracy and precision of these sensors.

The range of detection spans between 10^{-5} and 10^{-2} M and the response is between 1 and 5 min, allowing up to 30 analyses every hour. The solution must be very weakly buffered in order to prevent loss of sensitivity.

The following list shows some of the reactions used by potentiometric biosensors.

1. H^+ cation:

$$\text{D} - \text{glucose} + O_2 \xrightarrow{\text{glucose oxidase}} \text{D} - \text{gluconolactone} + H_2O_2 \xrightarrow{H_2O} \text{D} - \text{gluconate} + H^+$$

$$\text{Penicillin} \xrightarrow{\text{penicillinase}} \text{Penicilloic acid} + H^+$$

$$\text{Urea} + H_2O + 2H^+ \xrightarrow{\text{urease@pH6.0}} 2NH_4^+ + CO_2$$

$$\text{Urea} + 2H_2O \xrightarrow{\text{urease@pH9.5}} 2NH_3 + HCO_3^- + +H^+$$

2. NH_4^+ cation:

$$\text{Urea} + 3H_2O \xrightarrow{\text{urease@pH7.5}} 2NH_4^+ + HCO_3^- + +OH^-$$

---

[1] $F = NA \cdot e$, $R = NA \cdot k$ where NA is the Avogadro Number (6.022×10^{23} mol^{-1})

$$\text{Creatinine} + H_2O \xrightarrow{\text{creatinine iminohydrolase@pH9}} N - \text{methylhydantoin} + NH_4^+$$

$$L - \text{asparagine} + H_2O \xrightarrow{L-\text{asparaginase}} L - \text{aspartate} + NH_4^+$$

In general:

$$L - \text{amino acid} + H_2O \xrightarrow{L-AA \text{ oxidase}} \text{keto acid} + NH_4^+$$

3. I^- anion:
   Detection of H_2O_2 from an oxidase reaction

$$H_2O_2 + 2H^+ + 2I^- \xrightarrow{\text{peroxidase}} I_2 + H_2O$$

4. CN^- anion:

$$\text{Amygdalin} + 2H_2O \xrightarrow{\beta-\text{Glucosidase}} 2 \text{ Glucose} + \text{Benzaldehyde} + H^+ + CN^-$$

Question 7

Explain the low accuracy of potentiometric biosensors using the Nernst equation.

3.6   Amperometric Biosensors

Amperometric biosensors detect a current due to redox reactions when a potential is applied at the working electrode (WE) against the potential of the electrolyte, which can be actively controlled by a potentiostat, or simply set to common ground, that is, 0 volt, by means of a much larger counter electrode (CE) of Type II. Such electrode is often called *quasi-reference electrode*. The response time generally corresponds to that of potentiometric biosensors; however, in this case the dynamic measuring range and the sensitivity are linearly dependent on the concentration of the redox-active species.

The Clark oxygen electrode (Fig. 3.19) forms the basis of most amperometric biosensors. It consists of a platinum cathode, at which oxygen is reduced, and an Ag/AgCl quasi-reference electrode (at 0 V).

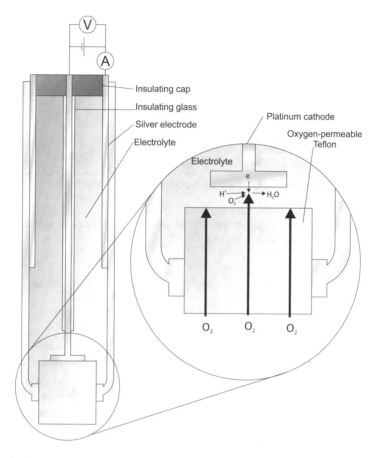

Fig. 3.19 A schematic representation the Clark oxygen electrode (Source: [4], © Larry O'Connell, CC-BY-SA 4.0 [5])

In order to measure the oxygen concentration as a function of the current flow, a bias voltage of − 0.6 V is applied at the platinum cathode vs. the Ag/AgCl anode. Both electrodes operate in a saturated KCl solution and are separated from the bulk solution by an oxygen-permeable membrane (e.g., Teflon).

The following reactions occur:

$$\text{Ag/AgCl anode} \quad 4Ag^0 + 4Cl^- \rightarrow 4AgCl + e^- \qquad \text{Pt cathode} \quad O_2 + 4H^+ + 4e^-$$
$$\rightarrow 2H_2O$$

Alternatively, hydrogen peroxide can be measured if the bias voltage at the Pt anode is set to +0.7 V vs. the Ag/AgCl cathode:

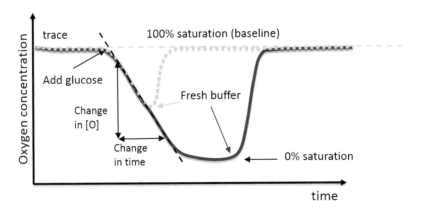

Fig. 3.20 Exemplary response characteristics of an amperometric biosensor with full-time (blue) and abbreviated interrogation (dashed yellow)

$$\text{Pt anode} \quad H_2O_2 \rightarrow O_2 + 2H^+ + 2e^- \qquad \text{Ag/AgCl cathode} \quad 2AgCl\ 2e^-$$
$$\rightarrow 2Ag^0 + 2Cl^-$$

The response signal versus time of an amperometric biosensor with glucose oxidase in the presence of a glucose solution is shown in Fig. 3.20. Between analyses, the biosensor is washed with a glucose-free, oxygen-saturated buffer.

The oxygen consumption runs at a steady rate. This is a convenient basis for obtaining reference response curves, for assessing unknown concentrations. Furthermore, only the initial portion of the response curve having a steady slope needs to be acquired and analyzed, thus shortening the analysis time (yellow dashed signal in Fig. 3.20). With this approach, the rinsing time for washing out the analyte while filling the cell with fresh buffer is also considerably reduced.

The performance of amperometric biosensors often depends on the actual concentration of dissolved oxygen. Mediators are able to transfer electrons directly from the enzyme to the electrode, bypassing the oxygen reduction, and thus addressing this issue. However, the mediators need to fulfill various requirements:

- Mediators should react quickly with the enzyme in its reduced form.
- Mediators should be appropriately soluble in both forms, that is, oxidized and reduced.
- Mediators should diffuse quickly from the enzyme to the electrode and vice versa, while remaining confined in the milieu of the biosensing layer; in other words, mediators should not go dispersed into the bulk solution.
- They should not be toxic.
- Regeneration of oxidized mediators should be pH-independent and work at low potential.

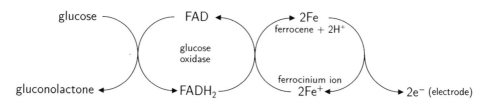

Fig. 3.21 Reaction scheme using ferrocenes as mediator. With *FAD* flavin adenine dinucleotide, *FADH₂* reduced form of FAD. This is a redox cofactor of the enzyme, that is, glucose oxidase (GOD or GOx) uses FAD to transfer electrons

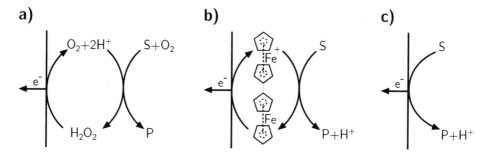

Fig. 3.22 Amperometric detection of the first (**a**), second (**b**), and third (**c**) generation

- Mediators in reduced form should not react with oxygen.

  Ferrocenes are a group of frequently adopted mediators. Figure 3.21 shows their reaction scheme.

  Three generations of amperometric biosensors have been developed so far (Fig. 3.22):

- Electrodes of the first generation measure directly H_2O_2 generated by the reaction itself ($E_0 \approx +0.7$ V vs. Ag/AgCl cathode).
- Electrodes of the second generation use mediators (ferrocenes) to transfer the electrons resulting from the catalytic activity ($E_0 \approx +0.2$ V vs. Ag/AgCl).
- Electrodes of the third generation provide direct electrons transfer from the reaction site ($E_0 \approx +0.1$ V vs. Ag/AgCl).

  The following reaction takes place in biosensors of the later type:

$$FADH_2\text{-Oxidase} \rightarrow FAD\text{-Oxidase} + 2H^+ + 2e^-$$

Examples of third-generation electrodes for amperometric biosensors are shown in Fig. 3.23.

Lowering the bias-potential E_0 as in the second and third generation has the big benefit of reducing the sensitivity for interferents. This can significantly ease the requirements for the protective/selective membrane.

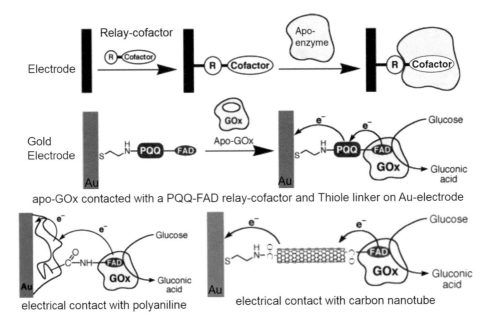

apo-GOx contacted with a PQQ-FAD relay-cofactor and Thiole linker on Au-electrode

electrical contact with polyaniline   electrical contact with carbon nanotube

Fig. 3.23 Direct transfer of electrons (plug-in electric contact) is achieved through a conductive crosslinker between the electrode and the enzyme cofactor. The following example indicates FAD as the cofactor of GOX and PQQ conducting the surplus of electrons out of the reduced enzyme to the electrode. A mediator is no longer necessary (Adapted from [6], © 2005, by permission from John Wiley and Sons)

Finally, we consider an application of amperometric sensors for detecting catecholamines [7, 8] (aromatic amines such as noradrenalin, dopamine, and adrenalin as well as their derivatives, which serve as messenger substances in the body). These hormones are produced, among other substances, by neurosecretory cells like neurons and chromaffin cells, and perform a variety of control functions over tissues and organs. The mechanism of catecholamine release at the cell membrane and the amperometric detection are shown in Fig. 3.24. In this approach, the cells work as bioreceptors if the aim is to investigate how external factors (e.g., drugs) can influence the physiology of exocytosis or recover normal function in pathologic states.

Question 8

Explain the functionality of amperometric biosensors and compare them with the functionality of potentiometric sensors.

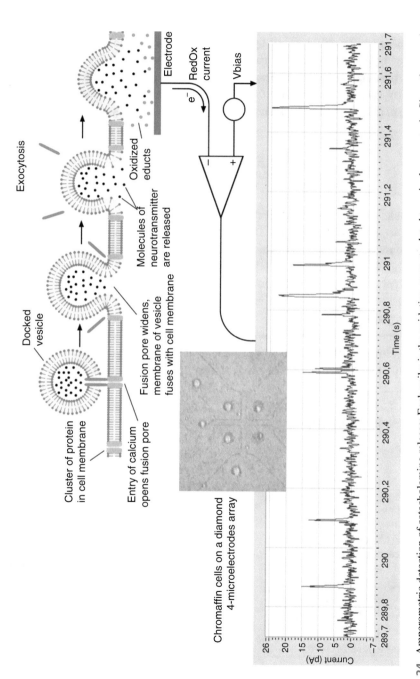

Fig. 3.24 Amperometric detection of catecholamine release: Each spike is the oxidation current related to a single exocytosis. Since the complete event has a very short duration of 10–50 ms, an enzyme-based detection is generally too slow and hence not effective for this application

3.7   Impedimetric Biosensors

The electrodes can be characterized by their impedance Z, which may change under the influence of both catalytic and noncatalytic biosensing activities. By applying a small sinusoidal potential $V(j\omega)$, one measures the current $I(j\omega)$ which flows through the electrode. By varying the frequency f over several decades, one gets the complex impedance $Z(j\omega)$ of the system, that is the sum of real and imaginary impedance parts, as function of ω:

$$Z(j\omega) = \frac{V(j\omega)}{I(j\omega)} = Z_r(j\omega) + jZ_i(j\omega) \quad \text{where:} \quad \omega = 2\pi f$$

The results of this Electrochemical Impedance Spectroscopy (EIS) measurement are usually fit to an equivalent circuit. The simplest yet realistic case, known as "Randles circuit," consists of a solution resistance (R_s), a double-layer capacity at the electrode surface (C_{dl}), a charge transfer resistance (R_{ct}), and a so-called *Warburg Element* (Z_w) having a typical 45° phase shift. However, factors like adsorption/desorption, enzyme binding, and electrode composition may lead to the need of more complex models for obtaining a good fit.

Z_w accounts for the diffusion of ions in an electrochemical reaction; thus, it appears at lowest frequencies (diffusional controlled region) since migration of molecules in solution is relatively slow. EIS measurement can be conveniently displayed in a Nyquist plot, here the frequency-independent R_s is given by the intercept with the real axis at the highest frequency. R_{ct} is the resistance against electron movement and is shown as the width of the kinetically controlled region, by extrapolating a second intercept with the real axis (Fig. 3.25).

Fig. 3.25 Equivalent circuit and Nyquist plot of an electrode represented by the Randles circuit model (Adapted from [9] © 2015 IntechOpen, CC-BY 3.0 [5])

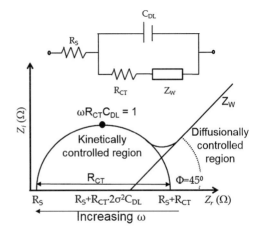

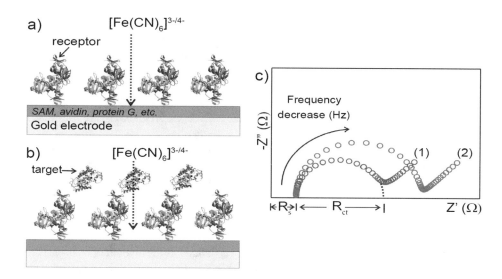

Fig. 3.26 Schematic representation of an impedimetric approach: (**a**) A bioreceptor is immobilized on a surface-modified gold electrode. A reversible redox couple like $[Fe(CN)^6]^{3-/4-}$ is used for transferring electrons under AC-biasing between the working electrode and the counter electrode, thereby the biolayer represents an impedance component, represented by R_{ct}. The EIS of the blank sensor is shown in (**c**) by the red data points. (**b**) When target analyte molecules bind to the receptors, the diffusion of the redox couple is further hindered, resulting in an increase of the impedance, as shown in (**c**) by the green dots. (Adapted form [11], © 2014 by the Authors, CC-BY [5])

In the following example of an impedimetric device, the bioreceptors are immobilized onto a gold electrode. A redox mediator like $[Fe(CN)_6]^{3-/4-}$ is used to exchange electrons and is faced with a barrier, quantified by R_{ct}, created by the receptive film, any bound target and the supporting chemistry. If target analytes are recruited from the solution (receptor–target interaction), the additional layer formed at the surface further increases the blocking effect and R_{ct}. Change in R_{ct} can thus be used as a direct quantitative reporter of specific target presence [10] (Fig. 3.26).

This latter method was shortly introduced here due to its electrochemical nature, although no catalytic activity is involved. However, the observed receptor–target binding interaction is a convenient preamble for introducing the next chapter on affinity biosensors.

Answer 1

Enzymes catalyze and therefore speed up a chemical reaction. This is because an enzyme lowers the energy barrier between the initial state (metastable) and the final state of an exothermic reaction, that is, a reaction with a negative variation of the free energy ΔG.

Answer 2

The Michaelis–Menten equation describes the velocity of an enzymatic reaction as a function of the substrate concentration. Here V_{max} and K_m are specific parameters of the reaction under observation.

Answer 3

Inhibitors reduce the velocity of a reaction for a given substrate concentration. This can happen most likely by increasing the Michaelis constant K_m (competitive inhibition) or by reducing the maximum velocity V_{max} (noncompetitive inhibition).

Answer 4

The international unit for enzyme activity (IU) is defined as the amount of enzymes that catalyzes 1 µM of substrate per minute, under standard conditions of temperature, optimal pH, and substrate concentration.

Answer 5

The sensitivity and the measuring range of the calorimetric biosensors are both very limited in most applications. However, it is possible to increase the sensitivity by cascading several reactions in the same pathway, thus increasing the cumulative heat output. In particular, the use of kinases, if applicable, can make available a large energy budget by binding/unbinding phosphate groups to/from the substrates. Reactions of this kind are called phosphorylation and dephosphorylation respectively.

Answer 6

Type 0 electrodes are made of chemically inert, conductive materials; therefore, they **do not** exchange ions with the electrolyte. Their function is to give/take electrons to/from the species undergoing reduction/oxidation at their surface respectively. The current direction, that is a reaction involving reduction or oxidation, is determined by the specific species involved in the process and the potential at which the electrode is biased (polarized).

Type II electrodes are fabricated with a semi-noble metal core covered with one of its hardly soluble salts. These electrodes are non-polarizable, that is they respond with a fairly linear current to the applied potential (Ohmic behavior). This is obtained by exchanging ions with the electrolyte in response to the voltage biasing. This means that the electrodes donate or gain mass during their use. If the salt layer is consumed it must be regenerated, otherwise the blank metallic electrode would behave similarly to one of Type 0.

Answer 7

The Nernst equation predicts the electrochemical potential, as a function of the concentrations of reduced and oxidized species. Yet this is a logarithmic relationship; therefore it is possible to measure the potential over a very wide range of concentrations, but this results in low resolution and accuracy.

Answer 8

Amperometric biosensors detect the target species by reduction or oxidation reactions. Such reactions can happen only if the working electrode is biased at the proper potential.

In potentiometric biosensors, the reactions leading to a potential change happen by themselves, that is, without the active help of the electrodes.

Take-Home Messages
- Catalysis speeds up chemical reactions. The effect can be of several order of magnitude.
- Enzymes are biological catalysts.
- The velocity of an enzymatic reaction depends on the formation and breakdown constants of a specific process and the concentration of the substrate. This is represented by the Michaelis constant and the Michaelis–Menten equation.
- Inhibitors reduce the velocity enzymatic reactions. It is therefore important to limit their presence in the samples to be analyzed as much as possible.
- Enzymatic reactions involve several chemical and physical processes, which lead to different detection strategies, like calorimetry, amperometry, and potentiometry.
- Electrodes of Type 0 and Type II are the basic elements for electrochemical detection methods.
- Potentiometry measures the analyte concentration by detecting the potential change caused by a reaction.
- Amperometry measures the analyte concentration by reducing or oxidizing the relevant species.
- EIS is an electrochemical method which does not necessarily involve chemical reactions, but simply measures how the detector impedance changes when target analytes are bound to receptors.

References

1. Jeremy M, Tymoczko JL, Stryer L. Biochemistry. 5th ed. New York: W H Freeman; 2002.
2. Nguyen HH, Lee SH, Lee UJ, Fermin CD, Kim M. Immobilized enzymes in biosensor applications. Materials. 2019;12(1):121.
3. Chaplin MF, Bucke C. Enzyme technology Cambridge: Cambridge University Press; 1990.
4. O'Connell L. A schematic representation of Clark and Lyons 1962 invention, the oxygen electrode. 2016.
5. The Creative Commons copyright licenses. https://creativecommons.org/licenses/.
6. Reconstituted redox enzymes on electrodes: from fundamental understanding of electron transfer at functionalized electrode interfaces to biosensor and biofuel cell applications. In: Bioelectronics; 2005. p. 35–97.
7. Carabelli V, Gosso S, Marcantoni A, Xu Y, Colombo E, Gao Z, et al. Nanocrystalline diamond microelectrode arrays fabricated on sapphire technology for high-time resolution of quantal catecholamine secretion from chromaffin cells. Biosens Bioelectron. 2010;26(1):92–8.
8. Tomagra G, Franchino C, Pasquarelli A, Carbone E, Olivero P, Carabelli V, et al. Simultaneous multisite detection of quantal release from PC12 cells using micro graphitic-diamond multi electrode arrays. Biophys Chem. 2019;253:106241.
9. Sergi B-O, Naroa U, Natalia A, Andrey B. Impedimetric sensors for bacteria detection. In: Toonika R, editor. Biosensors—micro and nanoscale applications. Rijeka: IntechOpen; 2015.
10. Santos A, Davis JJ, Bueno P. Fundamentals and applications of Impedimetric and redox capacitive biosensors. J Anal Bioanal Tech. 2014;2015:5.
11. Santos A. Fundamentals and applications of impedimetric and redox capacitive biosensors. J Anal Bioanal Tech. 2014:S7–016.

Affinity Biosensors

4

Contents

© The Author(s), under exclusive license to Springer Nature Switzerland AG 2021
A. Pasquarelli, *Biosensors and Biochips*, Learning Materials in Biosciences,
https://doi.org/10.1007/978-3-030-76469-2_4

Keywords

Epitope · Antibody · Antigen · Aptamer · Heterogeneous assay · Homogeneous assay · Labels

What You Will Learn in This Chapter
In this chapter, you will learn about the possibilities of sensor technologies based on the affinity binding of matching molecules. You will get an overview of the immune system and its antigen recognition mechanism. We will discuss homogeneous and heterogeneous immunoassays as well as different techniques for the detection of specific molecules. You will be able to select the correct method for detecting such substances and to make statements about the accuracy and sensitivity of the respective measurement method.

In this chapter, we deal with some of the key components of the immune system (antibodies, antigens, epitope), highlighting the exploitation of immunoglobulin-G as bioreceptor in affinity biosensors. The examples of physical transduction within the immunoassay methodologies discussed in this chapter are based on labeling techniques (fluorescence, enzymes, radioisotopes). Affinity biosensing methods based on other physical transduction methods will be discussed in Chaps. 5 and 6.

4.1   Immunosensor

An immunosensor is a biosensor utilizing the highly specific *affinity binding* between a capturing molecule and its counter molecule (see Fig. 2.1b). Bioreceptors can be antibodies or antigens, DNA or RNA, or aptamers.

In this sensors class, the physical transduction may base on a label detection (fluorescence, luminescence, colorimetry, etc.); an optical detection (absorption, interference, plasmon resonance, etc.); an electric current or voltage (electrode, ion-sensitive field-effect transistor [ISFET], electrolyte–insulator–semiconductor [EIS]); or a mass-sensitive detection (e.g., deflection of a cantilever or frequency shift of a piezoelectric resonator).

4.2   The Immune System

The immune system is a body-wide system, consisting of cells and organs specialized in defending the body against threats by "foreign" agents (antigens). This happens when molecular characteristics (so-called *epitopes*) are detected, which exist on the surface of pathogens (viruses, fungi, bacteria, toxins, but also chemicals and drugs!). The immune system shows an enormous *diversity* and *specificity*, and can recognize millions of distinctive non-self-agents for years, sometimes for life, thanks to specialized "memory cells." It is capable of producing customized cells and molecules to fight these "non-self" that is, foreign substances.

4.2.1   Antigen and Epitope

Any agent able to activate an immune response is called an *antigen*. This can be a molecule, a virus, a bacterium, a fungus, etc. The *epitope* is the identifying structure of an antigen and protrudes from its surface. Therefore, it was formerly called the *surface determinant*. The tissues and cells from other individuals also act as antigens (with the exception from an identical twin, which carry identical self-markers). For example, the immune system recognizes transplanted tissues as non-self and rejects them (in these cases, there is a need for an *immunosuppressor*).

4.2.2   Antigen Recognition Mechanism and Antibody Secretion

The defense against antigens happens as follows (Fig. 4.1):

- A microbe or virus (an antigen) enters the organism.
- The *B-lymphocytes*, which circulate in the blood and the lymphatic system, recognize the foreign molecular structure (epitope of an antigen).
- The antigen-stimulated B-lymphocytes grow and become *lymphoblasts* (plasma cells that begin to produce specific antibodies for the detected antigens).

The antibody concentration in the organism increases and the antibodies begin to recognize the viruses and bind their epitopes with their own binding sites (*Fab fragment*; fab: fragment antigen binding). Ultimately, the virus surface is covered with antibodies

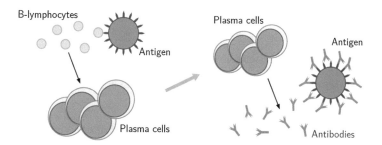

Fig. 4.1 (Left) Antigen recognition triggers the maturation of B-lymphocytes in plasma cells, which will produce and release specific antibodies for marking the invading pathogen (right)

Fig. 4.2 Destruction of a marked antigen by phagocytosis (left) or puncturing the cell wall (right)

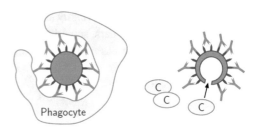

with the tails (*C-terminus*) facing outwards. These Fc fragments (Fc: fragment crystallizable) act as a marker for the body's natural destruction system.

The destruction of the "flagged" foreign bodies occurs in two ways:

- *Complement activation*: The complement system consists of up to 25 proteins divided in groups C1 to C9. These are sequentially activated to produce an inflammatory response and to control the activity of destroying pathogens by piercing the bacterial cell membrane (Fig. 4.2 right) and/or
- By *phagocytes*: Phagocytes are large white cells that can engulf and digest antigen–antibody complexes by *phagocytosis* (Fig. 4.2 left).

Antibodies are a family of large protein molecules known as immunoglobulins (Ig). Five families of human immunoglobulins have been identified:

- *IgG*: The predominant immunoglobulin in the blood. It works efficiently when covering microorganisms and is associated with the complement activation.
- *IgM*: First antibody available after infection and very effective due to its pentamerism.
- *IgA*: Concentrated in body fluids (tears, saliva, secretions of respiratory and gastrointestinal tracks).
- *IgD*: Antigen receptor of B cells and regulates their differentiation into plasma or memory cells.
- *IgE*: Familiar as the main character in inflammation and allergic reactions.

These five classes of immunoglobulins are schematically displayed in Fig. 4.3, together with a list of the most relevant parameters.

Question 1

Can you explain the concepts of antigen, antibody, and epitope?

The most common antibody used in biosensing is the immunoglobulin G. It is a complex protein with the typical Y-shape. The stem consists of two symmetric chains, also called *heavy chains*. The upper arms consist of a heavy chain and a light chain (each

The Five Immunoglobulin (Ig) Classes					
Properties	IgG monomer	IgM pentamer	Secretory IgA dimer	IgD monomer	IgE monomer
Structure	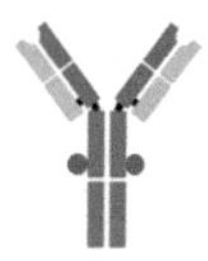		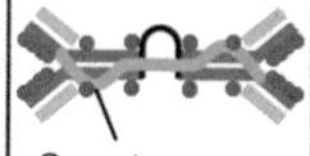 Secretory component	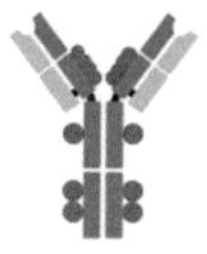	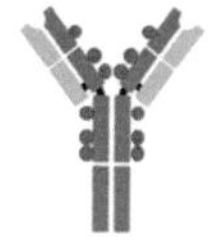
Heavy chains	γ	μ	α	δ	ε
Number of antigen-binding sites	2	10	4	2	2
Molecular weight (Daltons)	150,000	900,000	385,000	180,000	200,000
Percentage of total antibody in serum	80%	6%	13% (monomer)	<1%	<1%
Crosses placenta	yes	no	no	no	no
Fixes complement	yes	yes	no	no	no
Fc binds to	phagocytes				mast cells and basophils
Function	Neutralization, agglutination, complement activation, opsonization, and antibody-dependent cell-mediated cyotoxicity.	Neutralization, agglutination, and complement activation. The monomer form serves as the B-cell receptor.	Neutralization and trapping of pathogens in mucus.	B-cell receptor.	Activation of basophils and mast cells against parasites and allergens.

Fig. 4.3 The five classes of human immunoglobulin: The red circles symbolize sugar groups, that is the antibodies belong to the glycoproteins (Adapted from [1], © 1999–2018 Rice University, CC-BY 4.0 License. Sugar groups added [2])

arm is made up of a constant region and variable region). These variable regions make each antibody specific to an antigen (Fig. 4.4).

The antibody specificity stems from the fact that a molecular structure of a particular antigenic epitope fits into the binding side of the Fab part of an antibody like a hand in a glove or a key in a lock. Therefore, the variable domains of the Fab are different in different antibodies. In other words, the binding specificity of an antibody is based on the perfect complementarity of the three-dimensional crystallographic structure and the polar pattern of the epitope with the binding site of the antibody (Fab).

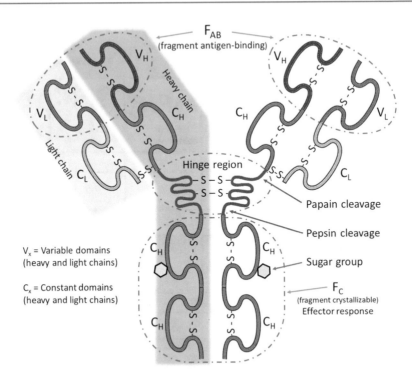

Fig. 4.4 Structural scheme of immunoglobulin G (the molecular mass is approx. 150,000 Da, the diameter 20–30 nm). Characteristic is the presence and position of disulfide bonds. Enzymes can cleave the antibody: Pepsin separates the Fc-terminus, leaving a twin Fab complex (connected at the hinge), which is called *Fab″*. Papain cleavage results in single Fab, called *Fab′*

4.2.3 Aptamers

The concept of aptamers (from Latin "*aptus*" – fit and Greek "*meros*" – part) was introduced independently by Szostak's and Gold's groups in 1990 [3, 4].

There are two types of aptamers:

- *Nucleic acid aptamers*: Most common aptamers are short, single-stranded DNA or RNA sequences (oligonucleotides, 15 to 60 bases long) engineered through repeated selection in vitro, to fit a variety of molecular targets like small molecules, proteins, nucleic acids, cells, tissues, and organisms (Fig. 4.5).
- *Peptide aptamers*: Artificially engineered to bind specific target molecules.

Typically, aptamers are developed with a process called Systematic Evolution of Ligands by EXponential enrichment (SELEX) [3], based on combinatorial chemistry, subsequent mutation, and further selection (Fig. 4.6).

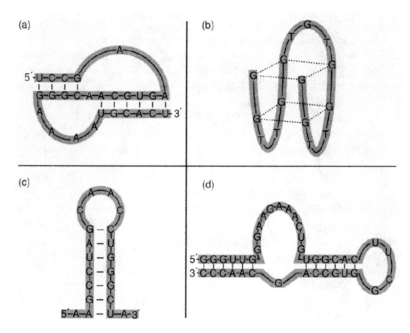

Fig. 4.5 Examples of aptamers: (**a**) RNA "pseudoknot" for HIV, (**b**) DNA "G-quartet" for thrombin, (**c**) RNA "hairpin" for polymerase, (**d**) RNA "stem-loop/bulge" for ATP (Reprinted with permission from [5], © 1995 American Chemical Society)

Both kinds of aptamers show binding affinity, selectivity, and specificity comparable to those of antibodies. Once properly "tuned," that is, when the optimal sequence is selected and known, aptamers can be replicated by direct synthesis. A comparison of Aptamens and Antibodies is shown in Fig 4.7 and Table 4.1.

4.3   Immunoassays

Immunoassays are test procedures in the field of biochemistry, which detect the presence and measure the concentration of a specific immunogenic analyte. The most typical arrangement of immunoassays consists of microtiter plates. There are several technological possibilities, such as 6, 24, 96 well plates; 384 well plates; and 1500 well plates. A microtiter plate can be analyzed by means of a plate reader (Fig. 4.8). The plate wells are functionalized at the bottom with antibodies; these are called more specifically *capture antibodies* (cAb), which selectively bind the target analyte, that is the *antigen* (Ag). A second *tracer* or *detection antibody* (dAb) is used in the so-called *sandwich assay* for detecting the cAb/Ag complex. This second antibody is marked with a label, which provides the signal for the detection in the plate reader. This label can be a fluorescent dye, a radioisotope, or an enzyme.

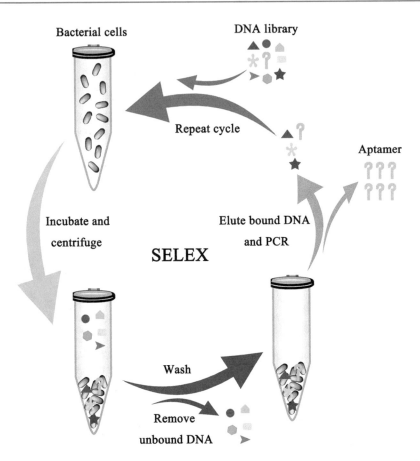

Fig. 4.6 SELEX process: Random sequence library undergoes partitioning for selection of binders for immobilized target; retained sequences are repetitively eluted and cycled through the selection and amplification (PCR) processes to isolate families of nucleic acid ligands for the target, in this case bacterial cells (Reprinted with permission of the Royal Society of Chemistry, from [6], © 2015)

Fig. 4.7 An estimated comparison of the size difference between an antibody (human IgG) and a selected aptamer (antithrombin DNA aptamer) is shown with space-filling models of both (Reprinted from [7], © 2006, with permission from Elsevier)

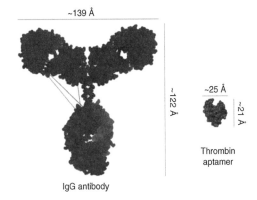

Table 4.1 Properties of aptamers vs. antibodies for their use in affinity biosensors [8]

Aptamers (oligonucleotides)	Antibodies (proteins)
Binding affinity is in low nanomolar to picomolar range	Binding affinity is in low nanomolar to picomolar range
Selection is an in vitro process that can target any small molecule, biopolymer, or cell	Selection requires a biological organism and is inefficient with toxins and small non-immunogenic molecules
Selection of aptamers is inexpensive and takes few weeks	Screening of monoclonal antibodies is expensive and time consuming
Uniform activity regardless of the batch	Activity of antibodies varies from batch to batch
Affinity parameters can be controlled on demand ("smart aptamers")	Difficult to modify affinity parameters
Wide variety of chemical modifications are introduced to diversify properties and functions	Very limited modifications
Recover original conformation after thermal shock	Temperature causes irreversible denaturation
Unlimited shelf-life	Limited shelf-life

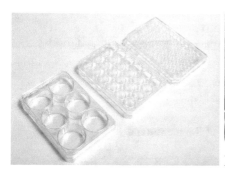 

Fig. 4.8 Immunoassays are commonly prepared in a microtiter plate: (Left) 6, 24, and 96 well plates. (Right) Plate reader for labels detection (Sources; left: [9] © Lilly_M, CC-BY-SA 3.0 (13); right: [10] © Pdcook, CC-BY-SA 3.0 [2])

The measurements carried out in the immunoassay protocols are influenced by the following factors:

- Selectivity, that is, the ability to discriminate between different substances
- Measuring range
- Accuracy
- Environmental condition (pH, temperature, ionic strength, etc.)
- Response time
- Recovery time

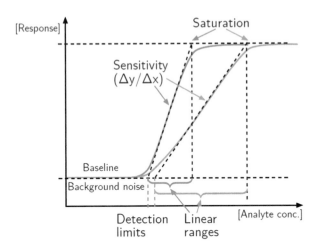

Fig. 4.9 Representation of measuring parameters: The baseline can be given either by the background noise or by initial conditions. Above the threshold of detection, we observe a response that grows until the sensor reaches saturation. The amplification can adjust the sensitivity. If saturated, the sensor cannot respond to a further increase of the concentration. The lower Limit of Detection is often indicated with its acronym *LOD*

- Working lifetime (and storage shelf time as well)
  The measuring range is characterized by the following parameters (Fig. 4.9):

- Linear range
- Detection limit: threshold or lower limit of sensitivity
- Saturation: upper limit of sensitivity
- Resolution: the minimum detectable variation of the measurand

  Further relevant parameters involved in experimental methods are as follows:

- Precision: Extent of random errors in the measurement.
- Accuracy: Deviation of measurement from the actual value.
- Repeatability: Variation arising when repeating the same measurement at exactly the same condition within a short time period.
- Reproducibility: Like repeatability but after longer periods of time or at different conditions, like locations, operators, and adopted instruments.

  Immunoassays can be classified as heterogeneous or homogeneous. In *heterogeneous immunoassays*, the components of the assay are added or removed sequentially, that is, *the measurement consists of several steps*. This method is suitable for both small and large molecules.

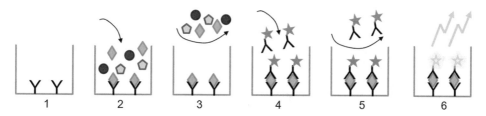

Fig. 4.10 Multiple steps for a heterogeneous immunoassay. The antigen is "sandwiched" between the capture and the tracer antibodies

In *homogeneous immunoassays*, all components of the assay are present at the same time in the liquid phase, that is, *the measurement consists of one step*. This method is best suited for the detection of small molecules, such as drugs, hormones, and peptides.

Traditional bioaffinity assays belong to the heterogeneous class. They are performed in several steps of reactions and washing phases for enhancing the assay specificity and for decreasing background interference caused by free unbound tracer molecules or labeled reagents within the reaction volume.

The heterogeneous *sandwich* assay is based on solid-phase binding (antibodies are immobilized on the bottom of the well) and requires the following steps (Fig. 4.10):

1. Immobilization of capturing antibodies at the bottom of the well.
2. Incubation of sample antigen with solid-phase bound capture antibodies.
3. Wash of sample – typically performed multiple times.
4. Incubation of labelled detection antibodies with capture antibody–antigen complexes.
5. Wash of excess unbound tracer.
6. Measurement of signal.

Question 2

Can you describe in written form the steps of a heterogeneous immunoassay?

4.3.1 Competitive Immunoassay

Another possibility is to adopt a competitive immunoassay protocol. Instead of a proportional toward analyte-concentration immunoassay, the competitive measurement takes place between a known reference concentration of labeled molecules and an unknown concentration of non-labeled analytes. The analyte competes with the labeled reference molecules and cleaves part of them.

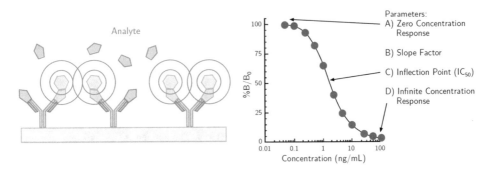

Fig. 4.11 (Left) Illustration representing a competitive immunoassay. (Right) The response is a sigmoid with negative slope and the maximum sensitivity (slope) is at the inflection point (IC_{50})

The reference concentration is usually calculated to "bias" the response in proximity of IC50 (Fig. 4.11) for maximum sensitivity, thus achieving an amplification for small concentrations of sample analytes. The amount of bound labels is inversely proportional to the analyte concentration. Errors in measurement are smaller near midscale and bigger at the extremities. This method has the important benefit of not needing a second labeled antibody (cost saving).

4.4 Assay Labeling

Based on the adopted labeling method for detection, immunoassays are classified as follows:

- *FIA–Fluorescence Immunoassay*: A fluorescent label is excited by light. The emitted light is measured.
- *EIA–Enzyme Immunoassay, ELISA – Enzyme-Linked Immuno-Sorbent Assay*: Enzyme label. The color change of a chromophore is analyzed by a photometer.
- *RIA–Radio Immunoassay*: Labeling with radioactive isotopes. The emitted radiation is detected by an X-ray film or a gamma camera.

4.4.1 Fluorescence Labeling

Fluorescence is the result of a three-step energetic process occurring in certain molecules called *fluorophores or fluorescent dyes*. The *Jablonski diagram* illustrates conveniently this process (Fig. 4.12):

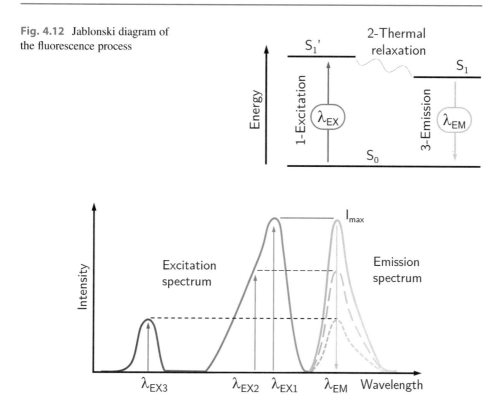

Fig. 4.12 Jablonski diagram of the fluorescence process

Fig. 4.13 Spectral diagram of fluorescence showing the Stokes shift. If one excites the fluorophore with a wavelength corresponding to the peak of its excitation spectrum, it will also yield an output at the maximum intensity. If the dyes are excited at a less convenient wavelength, the resulting output will have a lower emission intensity

1. Excitation of electronic state (from $S0$ to $S1'$).
2. Excited-State Lifetime (1–10 ns) with thermal relaxation (small energy loss).
3. Fluorescence Emission: $\lambda_{EM} - \lambda_{EX} = $ *Stokes Shift* (Fig. 4.13).

Consdering the Jablonski diagram shown in Fig. 4.12, the fluorescence quantum yield is defined as follows:

$$Quantum\ yield = \frac{Number\ of\ emitted\ photons\ (Step\ 3)}{Number\ of\ absorbed\ photons\ (Step\ 1)}$$

In order to detect fluorescence, the following instrumentation is required (Fig. 4.14):

1. Excitation source
2. Fluorophore

Fig. 4.14 Fluorescence detection. Light source, filter, and detector are usually integrated in the plate reader. The excitation is orders of magnitude more intense than the emission; therefore, the use of a blocking filter is crucial for assuring good sensitivity and dynamic range

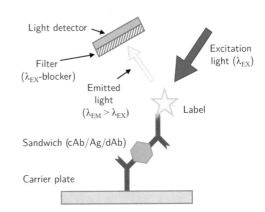

3. Wavelength filters
4. Detector

Different fluorescent substances have their specific absorption and emission wavelengths. Here are reported two examples of commonly used dyes in immunoassays:

- Fluorescein (FITC) is excited at 490 nm and emission is collected at 530 nm.
- Texas Red (TR) excitation at 595–605 nm, emission at 620 nm.

There are several commercial fluorophores that offer a wide variety of excitation and emission spectra, for example, the Alexa Fluor Dyes family [11], invented by Molecular Probes, now part of Thermo Fisher Scientific and distributed under the brand Invitrogen, which is particularly suited for labeling antibodies, covers a wide spectral range as shown in Table 4.2:

Fluorescent substances have a limited stability and may lose their effectiveness. *Photobleaching* is the irreversible photochemical degradation of the excited fluorophores under high intensity and/or repeated illumination conditions. This is due to formation of *reactive oxygen species* (ROS) under illumination, which not only cause degradation of fluorophores but can also produce illumination-related toxic effects called *phototoxicity*. It is therefore important to properly deal with photobleaching, either by adopting more photostable fluorophores or by adopting a smart approach for delivering excitation light selectively and in the needed amount, that is, fine-controlling exposure time, intensity, and location, as proposed for example by the CLEM method [12] (Fig. 4.15).

Question 3
Can you explain fluorescence and the Stokes shift?

Table 4.2 Absorption and emission wavelengths of the Alexa Fluor series of fluorophores (Source: Thermo Fisher Scientific, www.thermofisher.com)

Alexa fluor dye	Absorption max (nm)	Emission max (nm)
Alexa Fluor 350	346	442
Alexa Fluor 405	402	421
Alexa Fluor 430	434	539
Alexa Fluor 488	495	519
Alexa Fluor 514	518	540
Alexa Fluor 532	531	554
Alexa Fluor 546	556	573
Alexa Fluor 555	555	565
Alexa Fluor 568	578	603
Alexa Fluor 594	590	617
Alexa Fluor 610	612	628
Alexa Fluor 633	632	647
Alexa Fluor 635	633	647
Alexa Fluor 647	650	668
Alexa Fluor 660	663	690
Alexa Fluor 680	679	702
Alexa Fluor 700	702	723
Alexa Fluor 750	749	775
Alexa Fluor 790	782	805

To achieve good measurement results (high signal-to-noise ratio), it is important to comply with a few key points:

- Maximize the efficiency of the photon collection, for example, by confocal detection or dynamic focus adjustment.
- Tune the photodetector to optimize the sensitivity for both high dilutions near the detection limit and high concentration that can cause signal saturation.
- Minimize photobleaching by appropriate selection of the fluorescent marker and adjustment of the laser power as an excitation source (as in the CLEM method).
- Maximize the "background darkness" by specifically select a suitable nonfluorescent substrate to improve the detection threshold and dynamic range.

Modern microarray scanners offer surface sensitivities better than 1 fluorophore equivalent per μm^2, concentration detection limits in the picomolar range, and dynamic ranges of four orders of magnitude.

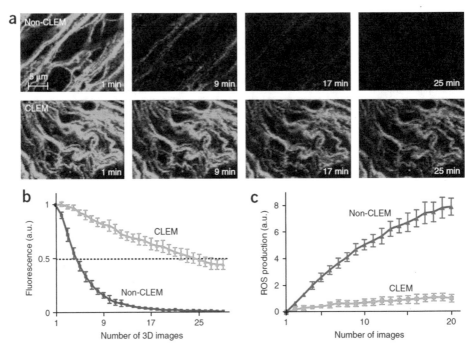

Fig. 4.15 (**a**) Photobleaching comparison between non-CLEM and CLEM observations of tobacco BY-2 cells expressing the green fluorescent protein GFP-MAP 4. (**b**) Photobleaching curves showing a sevenfold-reduced bleaching rate with CLEM. (**c**) ROS production is eightfold reduced by CLEM (Adapted from [12] © 2007, with permission from Springer Nature)

4.4.2   Enzyme Labeling

Enzyme Immunoassay (EIA) or Enzyme-Linked Immuno-Sorbent Assay (ELISA) is a heterogeneous measurement method widely used in the pharmaceutical and clinical environment [13]. In contrast to fluorescence labeling, it does not require an excitation source but rather a chromophore substrate (i.e., a pigment precursor) that is sensitive to the enzyme label (no spontaneous change by light, temperature, or impurities). The enzyme converts the substrate to the product and a color change occurs, which can be measured by a photometer (Fig. 4.16). The analyte concentration can then be determined by calculations. Here, the background can be compensated with a "blank" measurement (e.g., an *antigen-negative serum*, uses detection antibodies without enzyme labeling). The most commonly used enzymes are Horseradish Peroxidase (HRP) or Alkaline Phosphatase (AP).

The behavior of the system can be described by means of the law of mass action, which gives the Langmuir isotherm Γ. This describes the relative surface coverage by formation of a monolayer (Fig. 4.17).

The affinity association/dissociation of a receptor (R, antibody) and a ligand (L, antigen) can be described as follows:

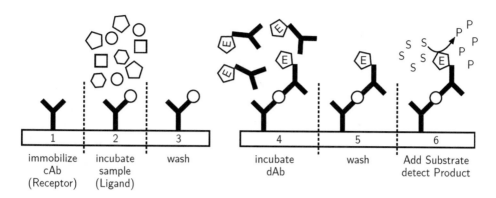

Fig. 4.16 ELISA sandwich. In this case, the label is an enzyme

Fig. 4.17 Langmuir isotherm Γ vs. the logarithm of ligand concentration: If K increases by higher affinity or optimized conditions, we can achieve the same surface coverage with a lower ligand concentration

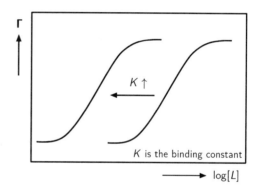

$$[R] + [L] \underset{k_d}{\overset{k_a}{\rightleftarrows}} [RL] \quad \text{where we define}: \quad K = \frac{k_a}{k_d}$$

On this basis, the Langmuir isotherm Γ is given by follows:

$$\Gamma = \frac{[RL]}{[R] + [RL]} = \frac{K[L]}{1 + K[L]} \quad \text{where} \quad [R] + [RL] = [R_{\text{tot}}]$$

In the above relationships, k_a and k_d are the constants of the kinetic reaction rate for association and dissociation, their ratio K is the so-called *binding constant*. [R] and [L] are respectively the concentration of the antibodies (receptors) and the antigens (ligands).

A selection of substrates typically used in ELISA protocols is shown in Table 4.3 and the results of an ELISA measurement performed in a microtiter plate is shown in Fig. 4.18.

Table 4.3 Overview of ELISA substrates

Enzyme	Substrate	Reaction	Max. absorption (nm)
Horseradish peroxidase (HP, HRP, HPO, or POD)	ABTS	Blue-green	405
	TMB	Blue	370 and 655
	Luminol	Luminescent	
Alkaline Phosphatase (AP)	BCIP	Blue	600–610
	BCIP	Red	460–505
	pNPP	Yellow	405–410
	Dioxetane complex	Luminescent	

TMB 3, 3', 5,5'-tetramethylbenzidine, *ABTS* 2,2'-azino-bis(3-ethylbenzothiazoline-6-sulfonic acid), *BCPI* 5-bromo-4-chloro-3-indolyl phosphate, *pNPP* = p-nitrophenylphosphatase

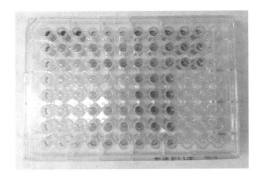

Fig. 4.18 ELISA microtiter plate reacted with HRP and TMB. Shown are analytes at different scaled concentrations (columns). Each reaction is performed in triplicate to minimize errors (rows). The rightmost column is the "blank" control (Source: Ajpolino, ELISA TMB, CC0 1.0 [2])

4.4.3 Chemiluminescence

Chemiluminescence occurs when a chemical reaction produces an electronically excited species that, upon emission of a photon, relaxes to an energetically lower state (possibly the ground state). The phenomenon occurs in nature, for example, in fireflies and several deep-sea organisms. It consists in the cleavage, in the presence of a catalyzer, usually a peroxide or enzyme, of an O–O bond or a phosphate group present in the chemiluminescent substrate.

$$A + B \overset{\text{Enzyme}}{\to} C^* + D$$

$$C^* \to C + h\nu$$

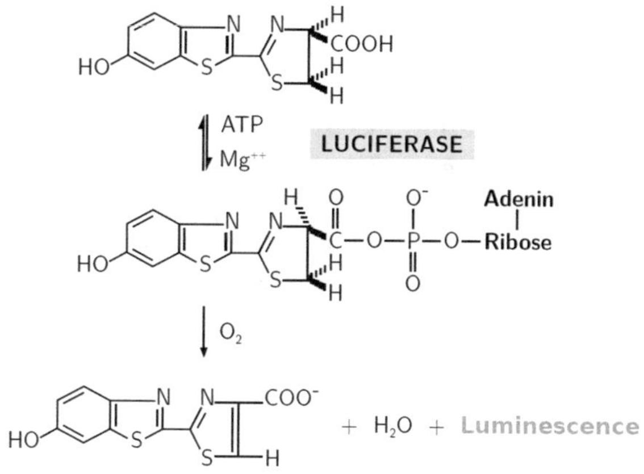

Fig. 4.19 Luciferin binds ATP and produces an excited molecule (bottom), which upon release of water and a luminescence photon, reverts to its initial state

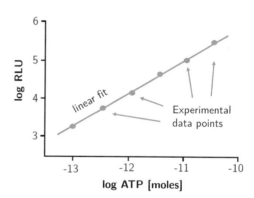

Fig. 4.20 ATP detection using luminescence (*RLU* relative luminescence units)

Chemiluminescence does not need a light source for excitation, shows no background signal or photobleaching, is highly sensitive (two orders of magnitude better than fluorescence), and reaches a dynamic range of 10^6 to 10^7. Luminescence detectors are films, photodiodes, PMT, and CCDs.

For example, the luciferin–luciferase reaction (bioluminescence) occurs in fireflies and requires the presence of magnesium, adenosine triphosphate (ATP), and oxygen [14] (Fig. 4.19).

This reaction is directly proportional to the ATP concentration. Thus, luminescence biosensors can be used to detect ATP, for example, for detecting bacterial infections, in the activated sludge analysis or for assessing food freshness and cleanliness of contact surfaces in cafeterias, kitchens, etc. [15] (Fig. 4.20).

4.5   Homogeneous Immunoassay Examples

In these immunoassays protocols, all components are present simultaneously in the liquid phase. It is necessary that such a measurement concept has an intrinsic selectivity between recognized (bound) and unrecognized analytes. In the first example that we analyze, donor and acceptor beads carry functional groups for a bioconjugation. When an affinity interaction occurs, a donor and an acceptor come in close proximity. This is, for instance, the principle of operation of the AlphaScreen® technology by PerkinElmer [16], based on the energy transfer from laser-excited oxygen molecules to luminescent complexes in the close vicinity. Upon laser excitation at 680 nm, a photosensitizer in the donor bead converts oxygen molecules from the ground triplet state $^3O2$ to a singlet excited state $^1O2$ (Fig. 4.21).

These excited oxygen molecules diffuse across to react with chemiluminescent molecules in the acceptor bead, which transfer at 340–350 nm these exciting energy to fluorophore dyes enclosed in the same bead. These fluorophores subsequently emit light at 520–620 nm. Since the lifetime of the excited state is only 4 µs, excited oxygen molecules can reach a maximum diffusion distance of 200 nm, which covers the distance of a bound acceptor bead, before they discharge the energy surplus. If no affinity binding takes place, the excited oxygen molecules cannot reach a far-floating acceptor bead in time and thus remain undetected. Note that the long laser wavelength of 680 nm cannot directly excite the fluorescence (Fig. 4.22).

Another example for homogeneous immunoassays makes use of the *Förster* (or *Fluorescence*) *Resonance Energy Transfer (FRET)* [17]. It is a distance-dependent $(1/d^6)$ interaction between electronic excited states of two label molecules in which

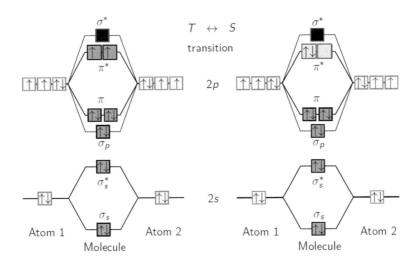

Fig. 4.21 Oxygen molecule: normal triplet state and transition to the singlet first excited state

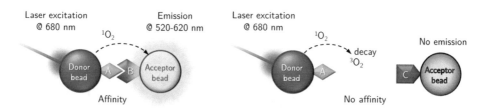

Fig. 4.22 Illustration showing the homogeneous immunoassay AlphaScreen® technology by PerkinElmer Inc. Only in case of affinity binding, the excited oxygen molecules can reach the acceptor bead and transfer its energy to the luminescent complex (left)

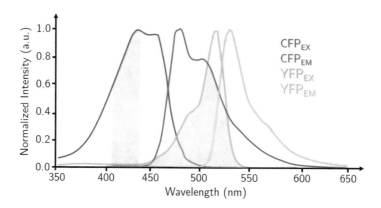

Fig. 4.23 Excitation and emission spectra of CFP and YFP. A resonant energy transfer is possible in the green-shaded overlapping region of CFP_{EM} and YFT_{EX}. Due to a partial overlap of CFP_{EX} and YFT_{EX} the primary excitation is suitable in the violet-shaded region only

excitation energy is transferred from a donor molecule to an acceptor molecule via dipole–dipole interaction, that is, without emission of a photon. The most common FRET fluorophore pair is cyan fluorescent protein (CFP) and yellow fluorescent protein (YFP) (Fig. 4.23), but several other pairs are available for fitting a variety of needs [18].

The basic conditions for FRET are:

- Donor and acceptor molecules must be in close proximity (typically 10–100 Å).
- The excitation spectrum (absorption) of the acceptor must overlap the fluorescence emission spectrum of the donor (Fig. 4.23).
- Donor and acceptor transition dipole orientation must be approximately parallel (Fig. 4.24).

Question 4

What is FRET? What are the conditions for it to happen?

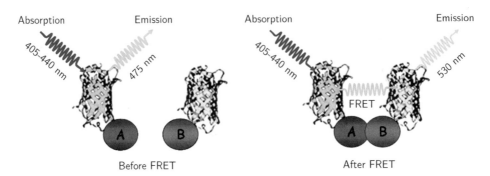

Fig. 4.24 FRET example – FRET between CFP as a donor fused to protein A and yellow YFP fused as an acceptor to protein B. Under favorable spatial and angular conditions, a dipole–dipole interaction between A and B causes a decrease in the intensity of donor (CFP) fluorescence concomitant with an increase in acceptor (YFP) fluorescence. In this very general illustration, CFP and YFP are depicted as cyan and yellow ribbon models fused to putative interacting proteins A and B, respectively (Adapted from [19], © 2006 Bhat et al., CC-BY-2.0 [2])

4.6   Radioimmunoassay (RIA)

Radioimmunoassay [20] is a relatively old method (>30 years). It has a very high sensitivity and reproducibility. Depending on the binding affinity of the target molecules, the isotopes ^{125}I, ^3H, ^{14}C, ^{32}P, ^{35}S are usually employed. They are weak gamma radiation sources. The half-life in case of ^{125}I is 60 days; therefore, waste handling is relatively unproblematic. The detection is done by counting the emitted gamma rays using a scintillator-based gamma camera or by using an X-ray film.

Although radioimmunoassay is the oldest immunoassay technique, it continues to offer distinct advantages in terms of simplicity and sensitivity. This technique requires complying with specific regulations for handling and disposal of radioactive substances.

The last example of this chapter shows a radioimmunoassay developed by PerkinElmer [16]. It is at the same time competitive and homogeneous with a radioisotope label. This radiometric proximity method uses microtiter plates coated with a scintillant material to enable the detection of specifically bonded radiolabeled molecules without the need for separation or a wash step.

The plates are functionalized with an anti-cAMP[1] antibody. This competitive assay uses ^{125}I-labeled cAMP as tracer. In the absence of cellular cAMP, the antibody binds the ^{125}I-labeled cAMP and brings it in close enough proximity to the scintillant on the plate, such that light is produced. In presence of cellular cAMP from a sample under test, it competes with the ^{125}I-labeled cAMP and thereby reduces the signal. In fact, γ-rays emitted

---

[1] Cyclic adenosine monophosphate (cAMP) is derived from ATP and belongs to the intracellular signaling system.

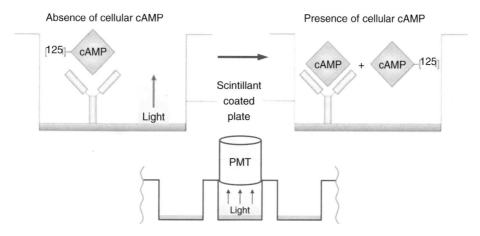

Fig. 4.25 FlashPlate® technology by PerkinElmer: (Top) Principle of operation (Adapted by permission from [21], Springer Nature © 2004). (Bottom) Read-out by photomultiplier tube

from decaying labels floating in the bulk have a low probability to interact with the scintillating layer at the bottom. The light signal is detected with a photomultiplier tube (PMT) (Fig. 4.25).

Question 5

Can you explain the difference between heterogeneous and homogeneous assays?

Question 6

Which forms of labeling did you learn so far?

Answer 1

Antigen is an external foreign entity (bacterium, virus, molecule, etc.) which enters a body and triggers an immune reaction.

Antibody is a protein produced by plasma cells in response to the penetration of an antigen. It is specifically tailored to fit the antigen and mark it, like a flag, as a target for the destruction mechanism.

Epitope is the characteristic morphology of the antigen surface, for example, the surface proteins. It is the epitope, which allows recognition of self- and non-self-entities, and it is on the epitope where the antibodies bind to the antigen.

Answer 2

A standard protocol of heterogeneous immunoassay consists of six steps:

1. Functionalization, that is, antibodies immobilization at the wells bottom in a microtiter plate.
2. Dispense the analyte solution and let it incubate.
3. Wash-out the analyte solution to remove all unbound material.
4. Dispense the solution with marker-conjugated antibodies and let incubate.
5. Wash-out to remove all unbound antibodies.
6. Detect labels.

Answer 3

Fluorescence is a light scattering process with partial energy degradation. After excitation with light of wavelength λ_{EX} the fluorophore molecule is in an excited energy state. Soon after it releases a fraction of the extra energy by thermal relaxation and finally emits the rest of the energy in radiative form by emitting a photon of wavelength $\lambda_{EM} > \lambda_{EX}$. The difference $\lambda_{EM} - \lambda_{EX}$ is called "Stokes shift."

Answer 4

It is a short distance interaction between two fluorescent molecules consisting in the transfer of energy from an excited donor to an acceptor fluorophore, which then emits at its longer wavelength. This is only possible if:

1. Donor and acceptor are in close proximity distance smaller than 10 nm.
2. The acceptor excitation spectrum must overlap the donor emission spectrum.
3. Since it is a dipole–dipole interaction, donor and acceptor dipoles must be approximately parallel to each other.

Answer 5

We have seen in question two that a heterogeneous immunoassay consists of several steps. Homogeneous immunoassays on the contrary have only one dispensing step before the readout. It means that washing-out for removing unbound analytes and marker-conjugated antibodies is not needed. Why is this possible? This is the result of "smart" approaches having an intrinsically selective read-out, which is sensitive only in the close vicinity of capturing molecules.

Answer 6

So far, we treated three families of labels:

1. Fluorophores, which absorb light in a specific spectral range and then fluoresce at a longer wavelength.
2. Enzymes, which turn a color precursor (chromophore) into a color or catalyze bioluminescence.
3. Radioactive isotopes. When they decay there is a radiation emission, typically a γ-ray.

Take-Home Messages

- Affinity sensors rely on the selective and specific binding properties of capturing receptors.
- In the immune system, affinity is provided by antibodies which bind antigens at their epitope. This is the primary reference model for developing affinity biosensors.
- Aptamers are becoming an interesting alternative to antibodies. They are developed in vitro with the SELEX process at much lower costs and ethical issues in comparison with antibodies.
- The protocol of heterogeneous immunoassays consists of several but simple steps (dispensing, incubation, wash, labeling, wash, readout).
- Homogeneous assays are prepared in one dispensing step plus readout, but require an intrinsically selective principle of operation for discriminating between bound and non-bound analyte.
- Labels can be classified in three groups: fluorophores, enzymes, and radioisotopes. Correspondingly there exist specific detection techniques and instruments for the read-out.

References

1. Yoo YK, Chae M-S, Kang JY, Kim TS, Hwang KS, Lee JH. Multifunctionalized cantilever systems for electronic nose applications. Anal Chem. 2012;84(19):8240–5.
2. The Creative Commons copyright licenses. https://creativecommons.org/licenses/.
3. Tuerk C, Gold L. Systematic evolution of ligands by exponential enrichment: RNA ligands to bacteriophage T4 DNA polymerase. Science. 1990;249(4968):505–10.
4. Ellington AD, Szostak JW. In vitro selection of RNA molecules that bind specific ligands. Nature. 1990;346(6287):818–22.
5. McGown LB, Joseph MJ, Pitner JB, Vonk GP, Linn CP. The nucleic acid ligand. A new tool for molecular recognition. Anal Chem. 1995;67(21):663A–8A.

6. Zhang C, Lv X, Han X, Man Y, Saeed Y, Qing H, et al. Whole-cell based aptamer selection for selective capture of microorganisms using microfluidic devices. Anal Methods. 2015;7 (15):6339–45.

7. Lee JF, Stovall GM, Ellington AD. Aptamer therapeutics advance. Curr Opin Chem Biol. 2006;10(3):282–9.

8. Nimjee SM, Rusconi CP, Sullenger BA. Aptamers: an emerging class of therapeutics. Annu Rev Med. 2005;56:555–83.

9. Lilly M. Multiwell cell culture plates-set. https://commons.wikimedia.org/wiki/File:Multiwell_cell_culture_plates-set.jpg.

10. Pdcook. BioTek Microplate reader. https://commons.wikimedia.org/wiki/File:Microplate_reader.jpg.

11. Alexa Fluor Dyes. https://www.thermofisher.com/de/de/home/brands/molecular-probes/key-molecular-probes-products/alexa-fluor/alexa-fluor-dyes-across-the-spectrum.html.

12. Hoebe RA, Van Oven CH, Gadella TWJ, Dhonukshe PB, Van Noorden CJF, Manders EMM. Controlled light-exposure microscopy reduces photobleaching and phototoxicity in fluorescence live-cell imaging. Nat Biotechnol. 2007;25(2):249–53.

13. Plested JS, Coull PA, Gidney MA. ELISA. Methods Mol Med. 2003;71:243–61.

14. Wood KV, Lam YA, McElroy WD. Introduction to beetle luciferases and their applications. J Biolumin Chemilumin. 1989;4(1):289–301.

15. Kricka LJ. Clinical and biochemical applications of luciferases and luciferins. Anal Biochem. 1988;175(1):14–21.

16. PerkinElmer webpage. https://www.perkinelmer.com.

17. Zhang X, Hu Y, Yang X, Tang Y, Han S, Kang A, et al. FÖrster resonance energy transfer (FRET)-based biosensors for biological applications. Biosens Bioelectron. 2019;138:111314.

18. Bajar BT, Wang ES, Zhang S, Lin MZ, Chu J. A guide to fluorescent protein FRET pairs. Sensors. 2016;16(9):1488.

19. Bhat RA, Lahaye T, Panstruga R. The visible touch: in planta visualization of protein-protein interactions by fluorophore-based methods. Plant Methods. 2006;2(1):12.

20. Dwenger A. Radioimmunoassay: an overview. J Clin Chem Clin Biochem. 1984;22(12):883–94.

21. Williams C. cAMP detection methods in HTS: selecting the best from the rest. Nat Rev Drug Discov. 2004;3(2):125–35.

Optical Biosensors

5

Contents

Keywords

Light sources · Optical components · Photodetectors · Absorption · Reflection ·
Refraction · Transmission · Total internal reflection · Evanescent waves TIRF

What You Will Learn in This Chapter

This chapter deals with optical biosensors. On the one the hand, we discuss the hardware (light sources, optical fibers, and waveguides as well as light detection methods); on the other hand, we deal with the physics of optical detection (absorption, refraction, reflection, evanescent waves) and the corresponding analysis methods like colorimetric sensors and total internal reflection fluorescence (TIRF).

In these sensing approaches, the biorecognition modifies or modulates how the light interacts with the sensing element, for instance, in terms of absorption when light passes through or is reflected/refracted, or even when marker labels are excited, for example, in a fluorescent detection scheme.

5.1 Basic Components

In the figure below, we see the basic components of a typical optical biosensor.

In the following, we analyze one by one the choice of components commonly adopted for providing the functions 2, 3, and 4 of Fig. 5.1.

5.1.1 Light Sources

Several technologies and related products are available as light sources. Most commonly utilized are lamps, lasers, and LEDs, whereas each group has own characteristics, advantages, and disadvantages.

The choice of the right light source for a certain use is based on the following selection criteria:

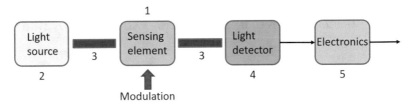

Fig. 5.1 Basic components of an optical system: (1) Chemical or biological sensing element provides the "modulation" of optical properties like absorbance, reflectance, etc., due to reaction or molecular binding in the element. (2) Light source for excitation of the sensing element. (3) Optical components (lenses, filters, polarizers, optical fibers, etc.) to couple light. (4) Light detector(s) to measure the light. (5) Electronic signal processing unit

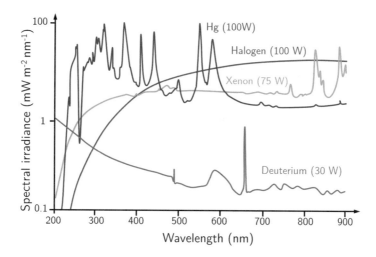

Fig. 5.2 Emission spectra for some lamp types

- Wavelength range
- Intensity range
- Coherence (laser)
- Modulation
- Stability (long-term drift, fluctuations)
- Power consumption
- Temperature/cooling
- Robustness

Lamps, like the tungsten filament, halogen and xenon lamps, and the mercury vapor lamps have a wide spectral emission, can reach high intensities, but require cooling. Mercury vapor lamps are often used in combination with spectral filters as a "line source", that is, only one of the many spectral peaks is selected. Xenon lamps have a uniform "white" light in the visible range, which extends into near UV, while Deuterium lamps have a remarkable ramping emission in the deep-UV range (Fig. 5.2).

Lasers as light sources offer many advantages: The light is monochromatic, stable, adjustable, and has coherent radiation, narrow beam cross-section, and high intensity. The choice of wavelengths is very wide, spanning from deep-UV to far-infrared (Fig. 5.3). Some of the most commonly used lasers are as follows:

- 457–514 nm (argon laser)
- 632.8 nm (HeNe laser)
- 375–1650 nm (semiconductor laser, laser diodes)

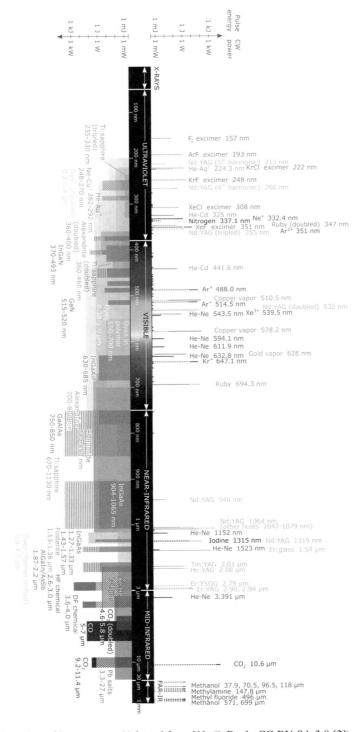

Fig. 5.3 Overview of laser sources (Adapted from [1], © Danh, CC-BY-SA 3.0 [2])

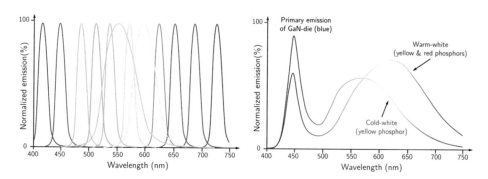

Fig. 5.4 Emission spectra of monochrome (left) and white LEDs (right)

Laser diodes are an affordable alternative to gas and crystal lasers, and suitable for many applications, however their emission is neither coherent nor strictly monochromatic.

LEDs have a narrow spectrum (typical *full width at half maximum* [FWHM] is 20 to 50 nm), but wider than lasers. Their emission ranges from a few milliwatts up to several watts. Their peak emission wavelength can be tuned by selecting the appropriate semiconductor composition and design parameters [3].

Their use is rapidly increasing due to the growing variety of available devices, in terms of power and wavelength. In several application areas, including among others fluorescence excitation, absorption spectroscopy, and plasmon excitation, LEDs are progressively replacing the traditionally used light sources due to lower cost, lower power consumption, intrinsic robustness, and longer lifetime.

Figure 5.4 shows spectra for monochrome and white LEDs. However, it should be noted that the definition of white light in LEDs is based on the sensitivity of the human eye and not on a homogeneous spectrum. In fact, the emission spectrum has a sharp blue component (primary emission of the gallium nitride die), which excites a coating layer that fluoresces in yellow. This combined emission is perceived as cold-white light, but the red and green spectral ranges are much less intense. More recently "warm-white" LEDs have been introduced. They have two fluorescent layers, yellow and red, which result in a wider spectral emission and a higher color-rendering index.

Question 1

Which light sources do you know for biosensors? Explain the respective advantages and disadvantages!

5.1.2 Optical Fibers and Waveguides

These components are primarily used for transmitting light along a certain pathway, if the simple transmission in air is not appropriate for a given instrumental setup. However, they

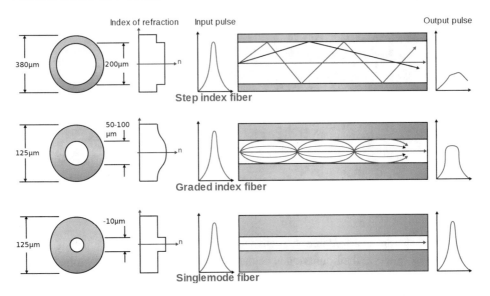

Fig. 5.5 Light transmission in a cylindrical optical fiber. Light enters and leaves the fiber at the two ends due to the nearly right angle. Along the travel, the light hits the sidewall at a flat angle, resulting in a total internal reflection, thus remaining confined inside the core. (Top, center) Multimodal transmission in fibers with step refractive index and graded index, the latter has a more uniform travel time. In a fiber with very narrow core, the light travels in single mode (bottom), yielding best timing characteristics and preserving light coherence when coupled to a laser source (Source [4] © Mrzeon, CC-BY-SA 3.0 [2])

can be used also as fundamental component in the sensing unit in applications exploiting the presence of an evanescent field close to their surface, as we will see later in this chapter.

The operation principle of optical fibers and waveguides is based on total internal reflection, due to the lower refractive index of the cladding with respect to the one of the fiber core. Depending on the cross-section of the fiber with respect to the wavelength and on the profile of the refractive index of the adopted material, it is possible to design optic fibers with different transmission modes as illustrated in the Fig. 5.5.

Optical fiber cables can be made out of plastic (inexpensive, robust, $\lambda > 450$ nm, maximum length of tens of meters) or glass (more expensive, $\lambda > 200$ nm, maximum length of kilometers is possible).

Planar waveguide plates can be produced by the means of bulk BK7 glass or crystals like sapphire, diamond, zinc selenide, silicon, and germanium, depending on the spectral range of use (Fig. 5.6 left). Microfabricated thin-film waveguides (also called "strip waveguides") are fabricated by means of the same technologies in use for micro- and optoelectronics (Fig. 5.6 right). In this case, other materials like gallium arsenide, gallium nitride, silicon oxide, silicon nitride, and polymers come additionally into play [5].

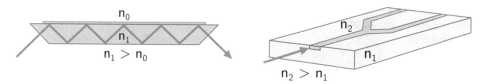

Fig. 5.6 Planar optical waveguides are available in form of plates (left) or as microfabricated integrated thin-film optics (right)

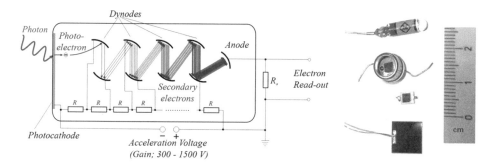

Fig. 5.7 (Left) Scheme of a photomultiplier tube; the photocathode has a spectral sensitivity of 100–800 nm. (Right) Photodiodes (Source [6], © Ulfbastel, CC-BY-SA 3.0 [2])

5.1.3   Optical Detectors

For detecting the modulated light, the choice of available sensors consists primarily of photodiodes (Fig. 5.7 right) for single-point measurements (typical sensitivity range for silicon-based devices is $\lambda \sim 350–1100$ nm) and charge-coupled device (CCD) or complementary metal oxide semiconductor (CMOS) image sensors (cameras) for acquiring full pictures (2D arrays) or light spectra (linear, 1D arrays). In case of single-spot measurements at very low-light intensity, photomultiplier tubes can represent a better option (Fig. 5.7 left).

5.2   Measurement of Optical Properties

The following optical properties are the variables examined during measurements with this class of sensors:

- Absorbance
- Chemiluminescence/bioluminescence emission
- Fluorescence/phosphorescence emission
- Refractive index
- Interference

Fig. 5.8 Absorption when light passes through a sample of thickness L

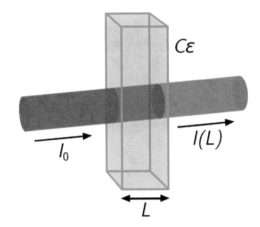

5.2.1 Absorption and Spectroscopy

When a sample, for example, a cuvette containing an analyte solution, is irradiated, the light intensity I progressively decreases as it passes through the sample. It is therefore a function of the pathlength L of the light through the absorptive substance. Such intensity decrease is due to *absorption*.

The absorption is formally described by the *Beer–Lambert law*:

$$I(L) = I_0 e^{-\varepsilon(\lambda)CL}$$

Here, $\varepsilon(\lambda)$ is the *molar absorptivity*, which is wavelength dependent, also called *extinction coefficient*, and C is the analyte concentration. The *absorbance* is defined as the logarithm of the intensity ratio between the initial intensity I_0 and the intensity at position L (Fig. 5.8):

$$A(\lambda) = \ln\left(\frac{I_0}{I(L)}\right) = \varepsilon(\lambda)CL$$

Hence, the absorbance is linearly dependent on the concentration. If a substrate or product of an enzyme-catalyzed reaction absorbs light, the activity of the enzyme can be quantitatively measured by monitoring the change of the substrate and/or product absorbance. For example, the natural redox-couple NAD^+/NADH (co-enzyme nicotinamide adenine dinucleotide) changes its absorption spectrum through oxidation and reduction. This substance is widely present in living cells and acts as a substrate for the enzymes of the oxidoreductases family. Its main role in metabolism is the transfer of electrons from one redox reaction to another.

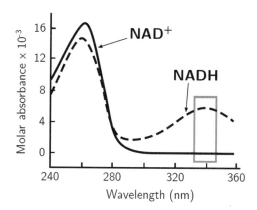

Fig. 5.9 Absorption spectrum of NAD$^+$/NADH

The comparison of the two spectra in Fig. 5.9 shows a large change of absorption at $\lambda \approx 340$ nm. Thus, this spectral region provides the highest sensitivity in the analysis of the reaction. Quantitative data regarding the enzyme activity can be extracted by using the Beer–Lambert law.

Question 2

How is enzyme activity determined by absorption spectrum of NAD$^+$/NADH?
In which region of the spectrum is the measurement most reasonable?

Another application is found in the instrument **pulse oximeter** (Fig. 5.10 left) for monitoring heart rate (HR), respiration rate (RR) and oxygen blood saturation (SpO$_2$) under different physiological (e.g., rest and activity) or pathological conditions. Here, red and infrared LEDs are used to illuminate a reasonably translucent area of the body with good blood flow. Typical adult/pediatric sites are finger, toe, and earlobe. Infant sites are the foot or palm of the hand and the big toe or thumb. Opposite to the emitters are photodiodes that receive the non-absorbed light. The oscillating waveform of the recorded signal is due to the pulsatile blood flow in the microvascular system and is called *photoplethysmogram* (PPG). From a biosensing viewpoint, we can consider hemoglobin as bioreceptor for oxygen. Its spectral absorption changes depending on the presence or absence of bound oxygen, that is, oxygenated versus oxygen-depleted status (Fig. 5.10 right). Furthermore, the signal oscillations allow calculating heart rate and respiration rate (Fig. 5.11).

At high oxygen saturation values (SpO$_2$ value close to 100%), which correspond to rest or healthy condition, the red signal is comparatively small, while the IR signal is relatively large. During physical exercise or respiratory deficit, SpO$_2$ values decrease strongly. Below 80% SpO$_2$, the red pulsatile signal becomes larger than the IR one.

The colorimetric test strips are another application. They are single-use cellulose pads impregnated with enzymes and reagents. Their common use is monitoring glucose and other bio-parameters in urine samples of diabetic patients (see Fig. 1.6 right). The pad for glucose detection is pre-filled with glucose oxidase (GOD), peroxidase (POD), and one

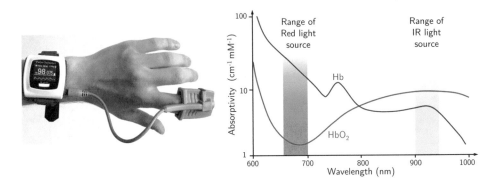

Fig. 5.10 (Left) Pulse oximeter (Source [7], CC0 1.0 [2]). (Right) Spectral absorption of oxygen-rich and oxygen-depleted hemoglobin, with the red and IR light sources used for the PPG

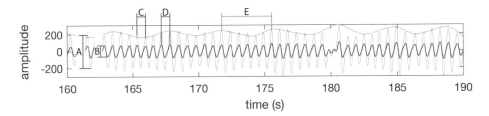

Fig. 5.11 PPG waveform features. (a, b) SpO$_2$ is calculated using the amplitude of DC and AC components of the IR and RED waveforms. (c) Single heartbeat in the IR waveform. (d) Single heartbeat in the red waveform. (e) Single respiration seen as a change in amplitude of the IR or red waveform. The green envelop wave is not part of the PPG, but is shown to highlight the respiration component of the PPG waveform (Reprinted from [8], © by the authors, CC-BY 4.0 [2])

chromophore. The H$_2$O$_2$, resulting from the oxidation of glucose with aerobic oxygen, oxidizes the chromogen to a colored dye.

$$Glucose + O_2 \xrightarrow{glucose\ oxidase} Gluconic\ acid + H_2O_2$$

$$Chromogen + H_2O_2 \xrightarrow{peroxidase} Color\ dye + 2H_2O$$

The evaluation is done by comparison with a color chart. To address color vision deficiency, which is quite common in diabetic patients, different color choices are available.

Question 3

How do colorimetry test strips work? When using a color strip, what is the light source? What is the light transmitting medium? What is the sensor, if not using an electronic instrument?

5.2.2 Immunoassays on Optical Fibers

By attaching a luminescence or fluorescence biosensor at the tip of an optical fiber, it is possible to both excite the sensor and transport the optical response signals away from the sensor site for a remote detection (Fig. 5.12 left). Used in conjunction with an endoscope or catheter, this instrumental setup allows direct in vivo measurements or measurements in hazardous or inaccessible environments, for example, in wells or deep waters.

Another option is to functionalize the tip of an optical fiber with immobilized antibodies. The fluorescence activity can then be detected. In this case, it is common to use a laser as light source. The returning fluorescence beam is guided to the detector via a dichroic beam splitter and possibly further electro-optical components: This method is particularly suitable for competitive assays, because the labels are preloaded before inserting probe into the measuring site and therefore it is not necessary to provide such markers and other reagents in the surrounding volume (Fig. 5.12 right).

5.3 Reflection, Transmission, Refraction

The angle of the incident light changes at the interface between two isotropic media with different optical densities, that is, refractive indexes (RI). By hitting the interface, a portion T of the incident beam is transmitted and refracted at an angle $\theta_t \neq \theta_i$, while the other portion R is reflected at an angle $\theta_r = \theta_i$ (Fig. 5.13). Under the assumption that there is no loss due to absorption, we have $T + R = 1$. Hereby, the refractive index n is defined as follows:

$$n = \frac{c}{v}$$

where c and v are the respective speeds of light in vacuum and in the medium.

Snell's law provides the relationship between the angles and the refractive indices of the two media:

$$\frac{\sin(\theta_i)}{\sin(\theta_t)} = \frac{v_i}{v_t} = \frac{n_2}{n_1}$$

where v_i is the speed of incident light in medium 1 of RI n_1, and v_t is the speed of transmitted and refracted light in medium 2 of RI n_2. The intensity of the reflected beam can be calculated by means of the Fresnel's laws. This calculation requires considering separately the fraction of the beam with S-polarization (TE-mode) and the fraction with P-polarization (TM-mode). For the two components, the reflection coefficients R_s and R_p are obtained by the respective equations:

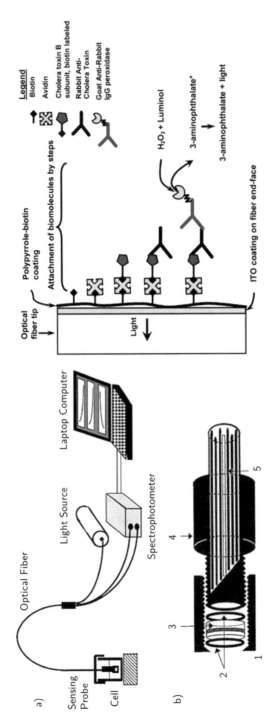

Fig. 5.12 (Left) Scheme of a portable fiber-optic biosensor (**a**) for detection of pesticides, for example, in river waters. Details of the sensor cell (**b**) with (1) screwable terminal holding ring, (2) O-ring's sealing, (3) catalytic biosensing sandwich, (4) probe tip, and (5) multielement optical fiber bundle (Adapted from [9], © 2002, with permission from Elsevier). (Right) Functionalization at the tip of a fiber optic chemiluminescent immunosensor for cholera detection (Reprinted from [10], © 2003, with permission of the American Chemical Society)

Fig. 5.13 Refraction and reflection at the interface between to media (here $n_2 > n_1$)

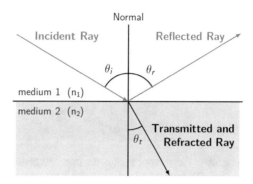

$$R_s = \left[\frac{n_1 \cos(\theta_i) - n_2 \cos(\theta_t)}{n_1 \cos(\theta_i) + n_2 \cos(\theta_t)} \right]^2 \quad \text{and} \quad R_p = \left[\frac{n_1 \cos(\theta_t) - n_2 \cos(\theta_i)}{n_1 \cos(\theta_t) + n_2 \cos(\theta_i)} \right]^2$$

At a certain angle, which depends on n_1 and n_2, R_p goes to zero and therefore the P-component is *totally refracted*, while the reflected beam is S-polarized. This angle is called *Brewster's angle*.

In the simple case of an incident beam orthogonal to the interface ($\theta_i = \theta_t = 0$), the above formulas reduce to

$$R_s = R_p = R = \left[\frac{n_1 - n_2}{n_1 + n_2} \right]^2$$

For example, at the air–glass transition with RI's of 1 and 1.5 respectively, $R = 0.04$ and $T = 0.96$, while in case of an air–diamond interface ($\text{RI}_{\text{diamond}} = 2.4$) we have $R = 0.17$ and hence $T = 0.83$.

5.3.1   Total Internal Reflection

When incident light travelling inside an optically denser medium ($\text{RI} = n_2$) hits the interface to a less dense medium ($\text{RI} = n_1 < n_2$) at an angle θ_i larger than the so-called *critical angle*, we observe a total internal reflection (TIR) at the medium boundary (Fig. 5.14). For calculating this angle θ_{crit}, we have the following formula:

$$\theta_{\text{crit}} = \arcsin\left(\frac{n_1}{n_2} \right)$$

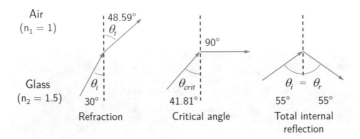

Fig. 5.14 Behavior of light at the interface between an optically denser and an optically less dense medium. The numerical values show, for example, the case of a glass-to-air interface. In this scheme, the partial internal reflection is not shown in the cases at left and in the middle. This internal total reflection is, for instance, the principle of operation of the light waveguide and the optical fiber

---

Question 4

Explain how an optical fiber works!

5.3.2 Evanescent Wave

If a wave hits a medium in which it cannot propagate, thus resulting in a total internal reflection, its amplitude behind the interface does not fall abruptly to zero. Instead, the amplitude decays exponentially and therefore the energy of the electromagnetic wave penetrates to a certain depth into the medium of lower refractive index. Such a wave is called *evanescent*. The depth of penetration d_p is defined as the distance at which the initial amplitude E_0 of the wave has dropped by a factor $1/e$, along the direction of penetration z (Fig. 5.15). This physical phenomenon is described by the following two formulas:

$$E_z = E_0 e^{-z/d_p} \quad \text{with} \quad d_p = \frac{\lambda}{2\pi\sqrt{n_2^2\sin^2\theta - n_1^2}}$$

For example, at a glass–water transition (with a glass having $n_2 = 1.6$, while water has $n_1 = 1.33$ and therefore $\theta_{crit} = 56.23°$) we obtain at angles between $57°$ and $90°$ values for d_p/λ between 0.894 and 0.179. Therefore, the local energy provided by the evanescent wave when TIR occurs, can be exploited and even conveniently adjusted for a selective investigation of just a very shallow layer at the boundary surface.

Question 5

How is defined the penetration depth of the evanescent wave in a medium?

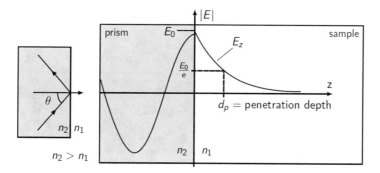

Fig. 5.15 Evanescent wave and the penetration depth in a total reflection

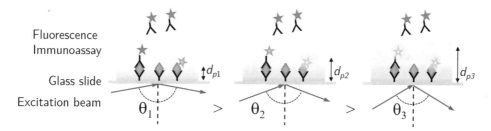

Fig. 5.16 The penetration depth of the evanescent wave can be adjusted with the angle, in order to provide the necessary selectivity of TIRF sensitivity, for example, when using sandwich or competitive assay protocols. This selectivity allows the development of homogeneous assays, because non-bound detection antibodies floating in the bulk cannot receive enough energy for exciting the labels

5.4   Total Internal Reflection Fluorescence (TIRF)

In total internal reflection fluorescence (TIRF), the evanescent wave is primarily responsible for electronic excitation of the fluorophore present in the lower refractive index medium. The penetration depth of the evanescent wave can be conveniently adjusted by changing the incident angle (Fig. 5.16). Due to the shallow penetration depth, only fluorophore molecules very close to the surface are excited, creating an extremely thin optical detection region. Outside the evanescent field, fluorescence is negligible. Thanks to this spatial selective property, this method is particularly suitable for homogeneous assays.

5.4.1   TIRF Microscopy (TIRFM)

Due to the shallow examination layer, the light collection in a microscope needs a careful optimization. In particular, the light pathway should be adjusted to generate the

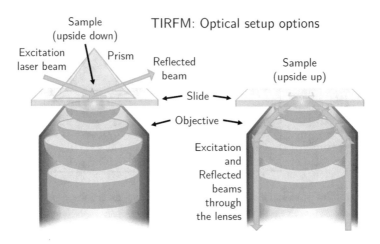

Fig. 5.17 Two optional schemes for a TIRFM instrument: Light excitation from above (left) and from below through the objective (right). Index matching oil is needed between the prism and the topside of the slide (left) and between the objective and the bottom side of the slide (right). For an easy coupling of the laser beam, the left scheme is more convenient, but the specimen is upside down, therefore it is not accessible for manipulation with external tools like electrodes and pipettes. For delivering the required buffers and other solutions, the design must include a fluidic cell. The right setup allows an easier handling of the specimen, but the objective requires a sufficiently large numerical aperture, and the microscope must have a hollow pathway for the laser beam. In both cases, the reflected beam is "discarded," while the weak fluorescent emission, collected by the objective, requires usually for its detection an Electron-Multiplying CCD (EMCCD) camera

fluorescence with the right efficiency. This includes the use of objectives with a large numerical aperture (NA [1]) as well as filling the gap between the front lens of the objective and the sample carrier glass slide with matching index oil to avoid an air–glass interface.

One easy option for configuring a TIRFM instrument is to add a prism and a laser light source to a standard microscope. The drawback of this design is that the sample must be placed upside down between the prism and the objective. A tailored flow-cell is usually adopted for addressing this problem (Fig. 5.17 left).

In another configuration, the laser beam reaches the sample by passing through the optics. This approach requires a lens with a numerical aperture larger than 1.4 for exceeding the critical angle (Fig. 5.17 right).

The next figure shows an example of simultaneous TIRFM and amperometric recordings of an exocytotic event from a bovine chromaffin cell (see for comparison Fig. 3.24). The cell was stained with NS510, a new fluorophore that accumulates in the vesicles and binds specifically to catecholamines. The TIRFM image highlights vesicles within approximately 100 nm from the bottom of the dish. Upon exocytosis, the

---

[1] NA $= n \sin \theta$, where n is the refractive index of the medium between the sample carrier and the objective and θ is the half-angle of the objective aperture.

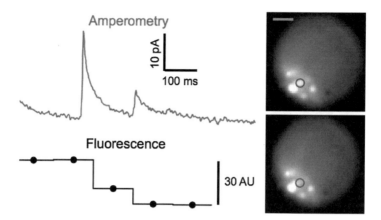

Fig. 5.18 Simultaneous TIRFM and amperometric detection of exocytosis from a bovine adrenal chromaffin cell (scale bar = 5 μm). Circles in the images depict the region of interest in which changes in NS510 fluorescence synchronous with amperometric spikes resulting from oxidation of catecholamine release from a single vesicle were observed (Adapted from [11], © 2019, with permission from John Wiley and Sons)

fluorophore is quenched, and the corresponding spot vanishes, thus indicating the location of the fused vesicle (Fig. 5.18).

5.4.2 TIRF Immunoassay

The time course of a TIRF immunoassay protocol is shown in Fig. 5.19. With this method, it is possible to determine the *equilibrium constants* of association and dissociation between receptor (e.g., antibody) and ligand (e.g., antigen). These values (K_a and K_d) are of great importance in many research areas, such as in the development of customized antibodies for therapeutic purposes. The yield of association, dissociation, and regeneration depends on the pH value. Association and dissociation use a buffer with physiologic pH, while the regeneration needs a buffer, which facilitates the ligand release by acting on the isoelectric points of receptor and ligand.

Next, we observe a competitive TIRF immunoassay method able to detect simultaneously two target analytes, and implemented on an optical fiber [11]. In this instrumental setup, the adopted **lock-in** synchronous detection allows rejection of artefacts and enhances the signal-to-noise ratio (Fig. 5.20).

The immobilized receptors are two *haptens*,[2] in particular a toxin (MC-LR) and an explosive (nitrobenzene, NB), both conjugated with ovalbumin as *carrier* protein. This

---

[2] Haptens are small molecules that cannot elicit an immune response by themselves. However, this becomes possible when they bind to a suitable protein called "carrier."

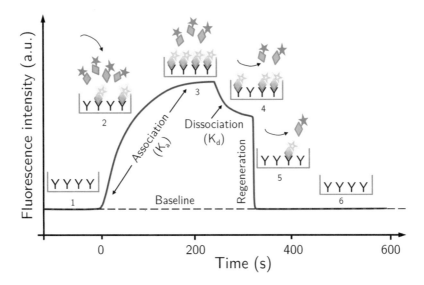

Fig. 5.19 Schematic structure of a TIRF sensogram. The protocol consists of the following steps: (1) Physiologic buffer running and achievement of a stable baseline; (2) injection of analyte (ligand) and increase of fluorescence signal due to association, until (3) reaching a quasi-plateau. At this point association and dissociation are in equilibrium. (4) Wash-out unbound analyte and decrease of fluorescence as a result of dissociation only. (5) A different buffer with appropriate pH forces the regeneration (i.e., the removal of all bound analyte), by leaving the receptors on the plate. (6) After reaching the baseline level, the plate is ready for the next measurement

conjugation provides the haptens with the requested antigenic behavior. Competition is observed when a nominal concentration of anti-MC-LR and anti-NB (this antibody also binds TNT) marked with two different fluorophores (Cy5 and Cy7, emitting at 667 and 776 nm, respectively) is mixed with a sample solution in which MC-LR and TNT concentrations are unknown, and this mix is delivered to the sensor. Antibodies bind competitively to both the immobilized receptors and to the target molecules (ligands) in the solution. The higher the concentration of the ligands in the sample, the lower is the free amount of antibodies available for binding to the immobilized receptors. Thus, the fluorescence intensity decreases with increasing concentration of the corresponding ligand. For the identification of the different analytes, two optical bandpass filters are available, which allow selecting the wavelength of the one or the other fluorescent marker.

Question 6

How can you conveniently exploit the spatial selectivity of TIRF?

Answer 1

We have basically three families of sources: Lamps, LEDs and LASERs.

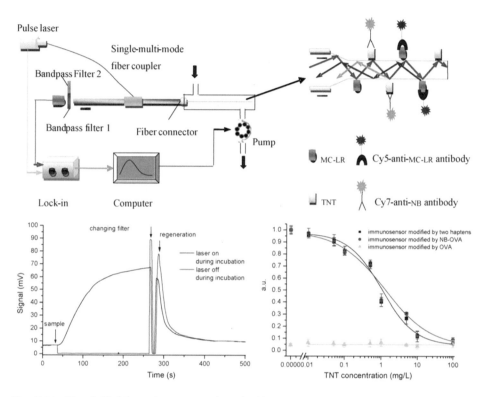

Fig. 5.20 (Top, left) Schematic representation of a fiber optic TIRF immunosensor. The laser light (modulated by a pulse generator) reaches the fiber tip carrying immobilized antibodies for the simultaneous detection of two small analytes, as shown at right. The fluorescent light is guided back by the optical fiber to a light detector, which is synchronous with the modulator (lock-in). The lock-in detection removes nonsynchronous events, that are due to external disturbances and artefacts, thus improving the signal-to-noise ratio. (Bottom, left) Sensogram recorded with and without laser excitation during association. Once the coefficients K_a and K_d are known for a specific receptor–ligand pair, it is possible to switch off the laser source at the end of the incubation time in order to prevent fluorophore bleaching. (Bottom, right) Calibration curve for TNT detection with three different functionalization schemes, both kinds of receptor (black), TNT competitor (NB-OVA) only (red), nonreceptive ovalbumin (green) only. (Adapted from [12], © 2010, with permission from Elsevier)

Lamps have a wide emission spectrum, typically from NIR to NUV, but with different profiles. For example, tungsten and halogen lamps emit most energy in the IR range and have a weak emission in UV, but they are inexpensive and very popular. Deuterium lamps have a remarkable emission in the UV range, while Hg-lamps have many sharp lines in the yellow to deep-UV range, which can be selected by means of spectral bandpass filters to obtain quasi-monochromatic light (e.g., for excitation of fluorescence). Most lamps release heat and its dissipation must be considered in the instrument design.

LEDs are becoming very popular due to the low cost, the high energetic efficiency, and the suitability for battery-operated and portable devices. Their emission spectrum is quite narrow. There is a wide choice of monochromatic LEDs distributed in a range from UV to IR. The die of short-wavelength LEDs (e.g., blue or violet) can be covered with fluorescent material to obtain a white light emission. However, the green and the red ranges are typically less intense than the blue and yellow ones.

LASERs are highly monochromatic and produce light beams with a very high energy density. "Genuine" LASERs emit coherent light, that is, all photons have the same phase, but are very expensive. LASER diodes are a good compromise in term of price–performance ratio if coherence is not needed.

Answer 2

The activity of enzymes having nicotinamide adenine dinucleotide as co-factor, can be measured by detecting the absorption of NAD^+/NADH in the UV range at 340 nm, which depends on the oxidation state. In fact, NAD^+ does not absorb at this wavelength, while NADH has a broad absorption peak.

Answer 3

Colorimetric strips have pads soaked with enzymes and chromophores. The oxygen needed for producing H_2O_2 is taken from the surrounding air. For example, the strip for glucose detection contains glucose oxidase (GOD), which in presence of glucose produces hydrogen peroxide. A second enzyme, peroxidase (POD), is activated by the presence of H_2O_2 and converts the chromophore in a color dye. The final color intensity is related to the glucose density in the sample. Here the light source is the ambient light, the transmitting medium is the air, while the light detector is the human eye. However, for a more precise assessment a spectrophotometer can be used instead.

Answer 4

Its principle of operation relies on the total internal reflection. In fact, the light travels within the optical fiber at a very flat angle with respect to the sidewall. The cladding has a lower refractive index than the core, which results in TIR along the fiber. At both ends the angle is nearly orthogonal, thus light can pass in and out.

Answer 5

The intensity of the evanescent wave is described by a falling exponential function.

The penetration depth d_p is defined as the distance at which the intensity E (z) decreased to 1/e of the initial intensity E_0. The resulting formula is $E(z) = E_0 \, e^{-z/dp}$.

Answer 6

TIRF is selective to the shallow penetration depth of the evanescent wave. This is in a first approximation a little smaller that the wavelength of the excitation beam. This selectivity can be used for instance in microscopy to selectively visualize structures close to the glass surface (TIRFM). Also very interesting is the suitability of TIRF for homogeneous immunoassays because non-bound markers will float in the bulk of the medium and cannot be excited, while markers associated to the recognized bound analyte are within the penetration depth of the evanescent waves and can be detected by fluorescence.

Take-Home Messages
- Optical properties and phenomena can be conveniently exploited for the physical transduction within biosensing methods.
- A typical read-out setup around the sensing element, requires a light source, a photo detector, and light-transmitting media, like lenses, optical fibers, etc.
- Measurement of absorption is convenient when the recognition of the analyte is directly related to, or can be translated by means of, cascaded reactions into a change of the molar absorptivity, for example, NAD^+/NADH, pulse oximetry, glucose oxidase–peroxidase (GOD-POD).
- At the interface between two media with different refractive indexes, the light is reflected and/or transmitted with refraction; the overall behavior depends on the angle of incidence.
- When light travels in the denser medium toward the less dense medium, we observe a total internal reflection (TIR) at the interface, for angles θ_i larger than the critical angle.
- TIR produces an evanescent wave at the opposite side of the interface.
- The short-range energy of the evanescent wave can be exploited for selective sensing of shallow layers in TIRF, TIRFM, and TIRF-immunoassays. It is also convenient for homogeneous immunoassays.
- The plot of the readout versus time is called Sensogram.
- Analyzing the sensogram of a TIRF-immunoassays allows the assessment of the equilibrium constants for association and dissociation of a given receptor–ligand pair.

References

1. Danh. Commercial laser lines. https://commons.wikimedia.org/wiki/File:Commercial_laser_lines.svg.
2. The Creative Commons copyright licenses. https://creativecommons.org/licenses/.
3. Winkler H, Vinh QT, Khanh TQ, Benker A, Bois C, Petry R, et al. LED components – principles of radiation generation and packaging. In: LED lighting. Weinheim: Wiley; 2014. p. 49–132.
4. Mrzeon. Optical fiber types. https://commons.wikimedia.org/wiki/File:Optical_fiber_types.svg.
5. SKS S. aP. Review on optical waveguides. In: You KY, editor. Emerging waveguide technology. Rijeka: IntechOpen; 2018.
6. Ulfbastel. Fotodio. https://commons.wikimedia.org/wiki/File:Fotodio.jpg.
7. UusiAjaja. Wrist-oximeter. https://commons.wikimedia.org/wiki/File:Wrist-oximeter.jpg.
8. Longmore SK, Lui GY, Naik G, Breen PP, Jalaludin B, Gargiulo GD. A comparison of reflective photoplethys mography for detection of heart rate, blood oxygen saturation, and respiration rate at various anatomical locations. Sensors. 2019;19:8.
9. Andreou VG, Clonis YD. A portable fiber-optic pesticide biosensor based on immobilized cholinesterase and sol–gel entrapped bromcresol purple for in-field use. Biosens Bioelectron. 2002;17(1):61–9.
10. Konry T, Novoa A, Cosnier S, Marks RS. Development of an "Electroptode" Immunosensor: indium tin oxide-coated optical fiber tips conjugated with an Electropolymerized thin film with conjugated cholera toxin B subunit. Anal Chem. 2003;75(11):2633–9.
11. Zhang L, Liu XA, Gillis KD, Glass TE. A high-affinity fluorescent sensor for catecholamine: application to monitoring norepinephrine exocytosis. Angew Chem Int Ed Engl. 2019;58 (23):7611–4.
12. Long F, He M, Zhu A, Song B, Sheng J, Shi H. Compact quantitative optic fiber-based immunoarray biosensor for rapid detection of small analytes. Biosens Bioelectron. 2010;26 (1):16–22.

Label-Free Biosensors

6

Contents

Keywords

Interferometry · Plasmon resonance · Cantilevers · Bulk acoustic waves · Surface acoustic waves

A. Pasquarelli, *Biosensors and Biochips*, Learning Materials in Biosciences,
https://doi.org/10.1007/978-3-030-76469-2_6

What You Will Learn in This Chapter
In this chapter, we focus on label-free biosensors. First, we look at sensors with optical detection. In this context, the Mach–Zehnder interferometer and the surface plasmon resonance offer the possibility to limit the optical examination to shallow surface layers. The second group are the gravimetric biosensors. The adsorption of molecules leads to an increase of the mass of a mechanical sensor like a cantilever or a crystal resonator. This change can be determined by measuring the bending of the cantilever or by measuring the change of the resonance frequency. You will be able to explain the underlying physical principles of the methods as well as to make statements about the measurement accuracy and the detection limits.

6.1   Mach–Zehnder Interferometer

The Mach–Zehnder interferometer (MZI) is a planar optical waveguide structure in which the light from a phase coherent laser is split into two parts and guided through two paths. One optical path presents a functionalized area directly on top of the waveguide core, where the affinity reaction can take place (so-called *sensing arm*), whereas the second path represents the reference beam (so-called *reference arm*) (Fig. 6.1).

The intensity of the output is given by

$$I = I_0 \cdot \frac{1}{2}\left[1 + \cos\left(\Delta\varphi\right)\right]$$

and the phase shift $\Delta\varphi$ between the two beams is given by

$$\Delta\varphi = \frac{2\pi}{\lambda} \cdot \Delta n \cdot L$$

Here Δn represents the change of the refractive index due to the analytes binding. This results from the interaction of the evanescent field on the surface of the optical waveguide with the analyte. For couplig the light into the MZI different techniques are available as shown in Fig. 6.2. A device exampe is displayed in Fig 6.3, which shows how the design parameters can influence the sensitivity of an MZI sensor.

The device response to the increasing protein layer thickness tends to saturate as the evanescent field region is filled by the biochemical layer (Fig. 6.4).

Fig. 6.1 Splitting and recombination of a laser beam within the Mach–Zehnder interferometer

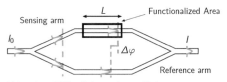

Two Y-junctions: 50% splitting and recombining

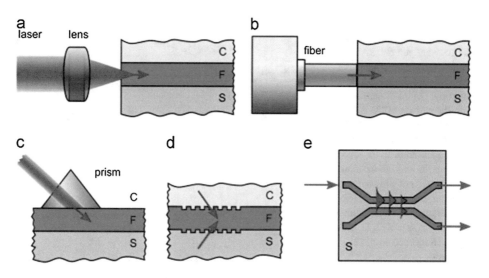

Fig. 6.2 Light coupling techniques for planar optical waveguides: (**a**) End-fire coupling; (**b**) butt-end coupling; (**c**) prism coupling; (**d**) grating coupling; (**e**) directional coupling (Reprinted from [1], © 2014, with permission from Elsevier)

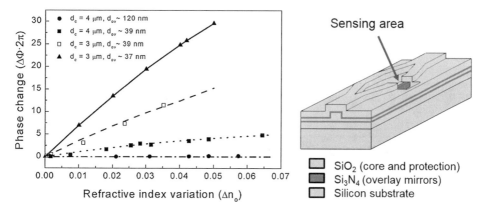

Fig. 6.3 Calibrating curves of an MZI-sensor with different waveguide structures: d_c is the SiO_2 core thickness with $n_c = 1.485$; d_{ov} is the Si_3N_4 overlay thickness with $n_{ov} = 2$. These mirror layers sandwich the core (Reprinted from [2], © 2003, with permission from Elsevier)

Question 1

How does a Mach–Zehnder interferometer work? What physical quantity is measured? How does the physical quantity correlate with the analyte being measured?

Fig. 6.4 Signal response of the sensor presented above to the antibody/antigen α-HSA/HSA immunoreaction. (**a**) Physical adsorption of the α-HSA antibody; (**b**) immunoreaction with the antigen HSA. Also in this case, like in the TIRF, the kinetics of the affinity reaction can be measured. In (**b**, top) the regeneration and the baseline recovery are not shown, that is, this sensogram is incomplete (*HSA* human serum albumin, the most common protein in human blood) (Reprinted from [2] with permission from Elsevier B.V.)

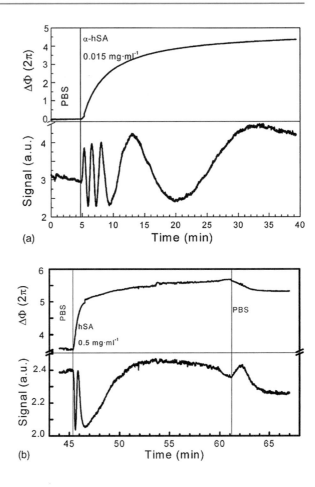

6.2   Surface Plasmon Resonance

A surface plasmon is a longitudinal charge density wave propagating along the interface between a dielectric and a metal. The electrons in the metal must exhibit free electron–like behavior. The evanescent field of the TIR can lead to the excitation of a surface plasmon resonance (SPR) as shown in Fig. 6.5. Among the metals with this property, gold is the most suitable one for use in an aqueous environment. The responses of copper and aluminum are too broad, furthermore those materials, same as silver, are susceptible to corrosion/oxidation.

In this context, the incoming beam is polarized in such a way that the axis of the magnetic field is parallel to the surface and that the electric field has components both parallel and perpendicular to the surface (Fig. 6.6). The resonance arises at the expense of the incident light energy, that is, the spectral component of light with the matching wave vector is transferred to the plasmon resonance. Therefore, it is missing in the reflected beam and the corresponding spot is dark.

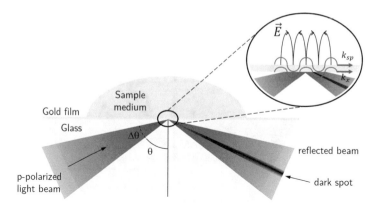

Fig. 6.5 Representation of the surface plasmon excitation: The incident beam is tapered in order to provide a range of angles $\Delta\theta$. In the reflected beam, there is a dark spot at a specific angle. This dark spot corresponds to the energy transferred to the plasmon resonance

Fig. 6.6 Field directions for the magnetic and the electric field of the surface plasmon resonance exciting radiation. Here ε_a is the dielectric constant of air and ε_m the one of the metal layer

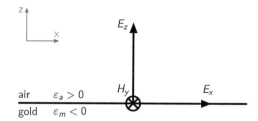

The dispersion relation represents the angular frequency of the electromagnetic field versus wave vector. Here below are the formulas for light in air (left) and for the surface plasmon in the metal layer (right):

$$\omega = c \cdot k_x \qquad\qquad \omega = c \cdot k_{sp} \cdot \sqrt{\frac{1}{\varepsilon_a} + \frac{1}{\varepsilon_m}}$$

ε_m is the dielectric function of the metal film. At resonance (SPR) ε_m is negative and light is absorbed.

Here we recall some useful formulas:

Wavevector in vacuum: $k = \frac{2\pi}{\lambda} \rightarrow \omega = c \cdot k$

Refractive index: $n = \sqrt{\varepsilon \cdot \mu} \approx \sqrt{\varepsilon}$

Velocity of light in medium: $v = \frac{c}{n} = \frac{c}{\sqrt{\varepsilon}} = \frac{\omega}{k}$

At all angular frequencies ω, the horizontal surface plasmon wave vector k_{sp} is greater than the horizontal component of the wave vector of the light in air k_x (Fig. 6.7) and therefore no resonance can occur (resonance needs the match of the two vectors k_{sp} and k_x). To reach a resonance, we need to increase kx. This can be obtained by tilting the straight

Fig. 6.7 Comparison of the
dispersion relations at the
interphase: for air ($\varepsilon_a = 1$) and
for surface plasmon in the metal.
The two dispersion relations do
not intersect and therefore no
resonance condition is possible
when light is directly coupled
from the air to the metal

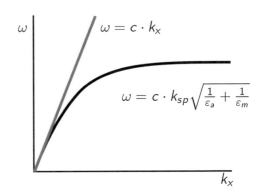

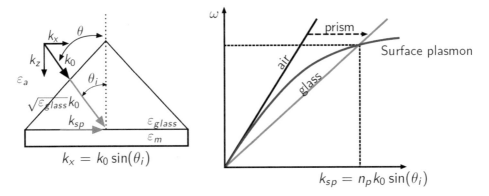

Fig. 6.8 (Left) the introduction of a prism increases the wavevector of the incident light. (Right) Dispersion relations for light in the air (black), in the glass (gray), and for surface plasmon at the interface (blue)

line of the dispersion relation of light in air in order to intersect the dispersion relation of the surface plasmon in the metal. This is done by adding a material with a dielectric function $\varepsilon > \varepsilon_a$ (e.g., glass), and thus increase the wave vector as shown in the following formula:

$$k_{x,\text{glass}} = \sqrt{\varepsilon_{\text{glass}}} \frac{\omega}{c} \cdot \sin(\theta_i)$$

where θ_i is the angle of incidence inside the prism (Fig. 6.8).

The refractive index of the prism is given by $n_p = \sqrt{\varepsilon_{\text{glass}}}$

For a certain combination of the angle of incidence θ_i and the wavelength λ of the incoming light, resonating excitation of surface plasmon can be achieved. Under this condition, the light energy is transferred to the SPR, and the reflection drops. When using monochromatic light for excitation the reflected beam presents a dark line at the resonance angle. Alternatively it is possible to excite the SPR with "white light" at a fixed angle, thus providing a wide range of wavelengths. In this case the spectrum of the reflected beam presents a deep at the resonant wavelength (Fig. 6.9).

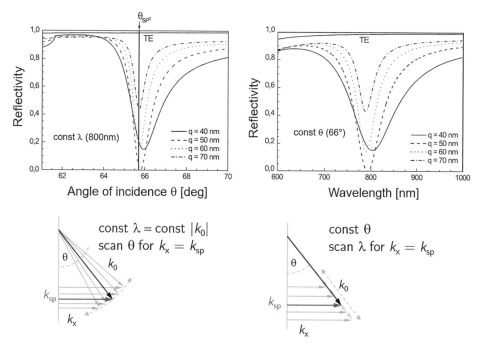

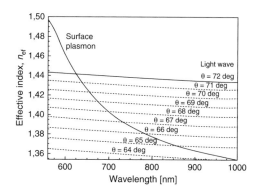

Fig. 6.9 (Top) Reflectivity vs. angle of incidence (left) and vs. wavelength (right) for different thicknesses of the gold layer q on borosilicate glass ($n \sim 1.52$) (Adapted by permission from [3] Springer-Verlag Berlin Heidelberg, © 2006). (Bottom) Vector representations for finding the resonance condition in the two respective configurations

Fig. 6.10 Resonance conditions for different values of the effective refractive index of the analyte in contact with a 50 nm gold layer deposited on an optical-grade borosilicate glass BK7 (Adapted by permission from [3] Springer-Verlag Berlin Heidelberg © 2006)

In biosensing applications, the gold film is typically covered with a water-based buffer having a refractive index (n or RI) moderately larger than the one of water ($n_{water} = 1.33$). Biomolecules immobilized over the gold film and bound analytes induce a variation of the local effective refractive index, which in turn influences the resonance condition. The experimental data for SPR as a function of the refractive index of the biosensing layer are shown in Fig. 6.10.

Question 2

Why, for surface plasmon resonance to occur, must the incident light beam strike the metal layer from the glass and not simply from the air?

6.3   SPR Affinity Biosensor

The SPR phenomenon can be utilized in the SPR affinity biosensor for detection of, for example, an affinity binding (Fig. 6.11).

The SPR measurement delivers a sensogram consisting of six characteristic segments (Fig. 6.12):

1. Initial baseline, only buffer solution flows.
2. Analyte is added and the association kinetics is measured.
3. Equilibrium between association and dissociation (K_A versus K_D).
4. Pure buffer is let flow and we can observe the dissociation kinetics (K_D only).
5. Regenerating solution is introduced to remove all analyte molecules from the immobilized receptors and thus regenerate the sensor surface.
6. Finally, pure buffer flows again and it is checked whether the output baseline has reached the same value of the initial baseline as a proof that the regeneration was effective and the conditions for a new measurement are correct.

The output signal provided by an SPR measurement is related to the analyte concentration according to the following relationship:

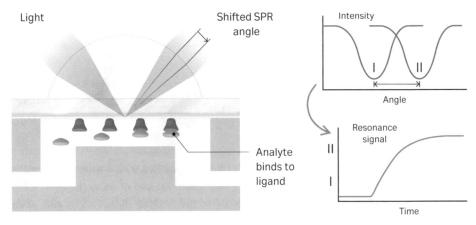

Fig. 6.11 (Left) Schematic representation of an SPR-affinity biosensor provided with a flow cell. (Right) Shift of resonant angle is recorded as function of time in a sensogram (© 2020 Cytiva— Reproduced by permission of the owner)

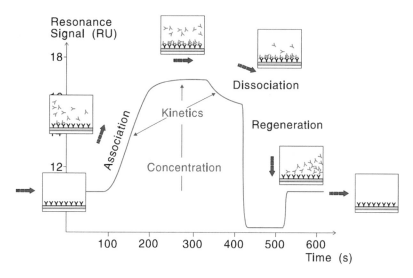

Fig. 6.12 Nominal sensogram of an SPR experiment showing the six characteristic segments described above. The unit RU is equivalent to 1×10^{-6} of the refractive index of vacuum. Distinctive of this plot, in comparison with the TIRF sensogram, is the signal level observed when the regeneration reaches a steady state, that is, when all ligands are removed and only the receptors are left on the chip surface. This is due to the refractive index of the regeneration solution, which is different from the one of the buffer (© 2020 Cytiva—Reproduced by permission of the owner)

$$S_c = \frac{\partial Y}{\partial C} = \frac{\partial Y}{\partial n_b} \frac{dn_b(C)}{dC} = S_{RI} S_{nc}$$

with Y resonance signal (angle or wavelength), C concentration of analyte, S_{RI}: sensitivity to refractive index profile change, S_{nc}: sensitivity to analyte binding (concentration c), dn_b: refractive index change due to analyte binding.

The refractive index change is related to the amount of proteins absorbed (g/cm²).

A practical rule of thumb is

$$\text{Change of } 1 \times 10^{-6} \, RI_{\text{vacuum}} = 1 \, \text{RU} = \tilde{1} \frac{\text{pg}}{\text{mm}^2} \text{ of adsorbed ligand}$$

The typical SPR limit of detection (LOD) for refractive index change is $\Delta n \approx 10^{-5}$ when measuring in the liquid phase and $\Delta n \approx 10^{-6}$ in the gas phase. LOD is affected by temperature fluctuations and noise (photodetector + amplifier + A/D converter).

The regeneration is a very important step in the SPR protocol, in fact it determines how reliably new experiments can be performed on the same chip, that is, without a "memory" of the previous measurements, but also with the same density and functionality of receptor molecules on the chip.

Antibodies are usually regenerated with low (<3) or high (>9) pH levels in solutions with low ionic strength (<30 mM). In general, antibodies against proteins are best regenerated using low pH, while antibodies against hapten–carrier complexes often respond better at high pH. Diluted buffer solutions of 10 mM glycine-HCl (low pH) or 10 mM glycine-NaOH (high pH) can be used to provide optimal pH values for different analysis needs.

6.4   SPR Systems

In Fig. 6.13 is displayed an example of a very basic SPR system for didactic purposes, implemented on the basis of a scheme described in [4]. For professional use, for example, in clinical diagnostics and drug-screening research in pharmaceutical companies, it is necessary to have very precise and reproducible working conditions, including among others temperature control; fine control of flow rates; automated handling of buffers, reagents, and analytes; and the corresponding software for managing the experiments and analyzing the data.

Several commercial companies offer SPR systems for different application scenarios and budgets. Among others, we find Cytiva with the Biacore family [5], FortéBio with the Pioneer systems [6], Reichert with the 2SPR and 4SPR systems [7], Brucker with the Sierra product line [8], Affinité with its open architecture P4SPR platform [9], and Plasmetrix specialized in affordable compact systems (e.g., for education) [10]. In the following, we give a closer look at the Biacore system.

6.4.1   Biacore™

One of the most representative SPR instruments is offered under the brand "Biacore" (now part of Cytiva). The plasmon excitation is provided by a convergent p-polarized light beam projected through a glass prism to the surface of a thin gold film (50 nm) evaporated on glass. The reflected light is measured by means of a CCD detector. The convergent

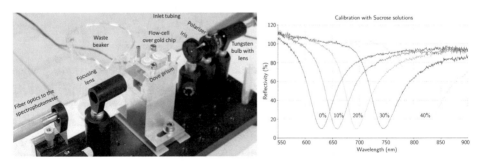

Fig. 6.13 (Left) Basic SPR setup for didactic experiments in the author's laboratory, measuring the spectral reflectivity. (Right) Calibration spectra with different sucrose solutions

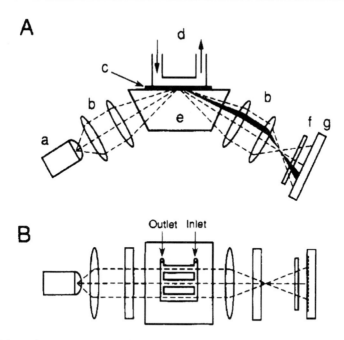

Fig. 6.14 Schematic representation of an angular modulation–based SPR sensor, like the early Biacore systems (**A** side view, **B** top view). (**a**) LED light source, (**b**) lenses, (**c**) sensor chip, (**d**) microfluidic cell, with three channels, (**e**) coupling prism, (**f**) polarizer, (**g**) detector array (Reprinted from [11], © 1991 with permission of the American Chemical Society)

fan-shaped light beam includes all angles of incidence, so that the resonant condition of the plasmon can be observed as a function of the angle without any moving parts. Biacore systems have been evolving over the decades, reaching at present, with eight twin parallel fluidic channels, a high degree of flexibility and performances (Figs. 6.14, 6.15, and 6.16).

6.5 SPR Detection

SPR detects changes of refractive index close to the *surface*. The refractive index unit (RIU) is the one of vacuum $= 1$; since the SPR method has a sensitivity of ppm or better, it is common to use a smaller unit, the RU which is equivalent to 1 μRIU. The accumulation of 1 pg/mm^2 gives a change of approximately 1 μRIU or 1 RU. Since all biomolecules have refractive properties, *no labeling* is required.

This results in the following:

- No need to separate bound from free (homogeneous method).
- Measurements are in *real-time*, thus allowing determination of *kinetic* data.
- SPR works with unmodified analyte (label-free).

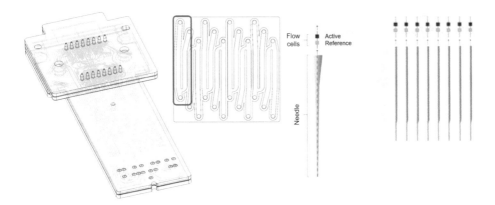

Fig. 6.15 (Left) Microfluidic cell of Biacore 8K+ system: By opening and closing the corresponding pneumatic valves, the analyte(s) can flow-in parallel or serially through the eight measurement channels having each two measuring spots (center), thereby allowing to perform complex measurement protocols. (Right) Analyte samples are injected by means of 8 needles (© 2020 Cytiva—Reproduced by permission of the owner)

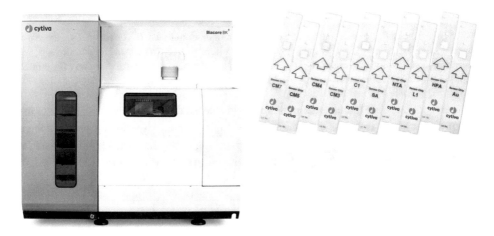

Fig. 6.16 (Left) Biacore 8 K+ system for high-end professional applications (e.g., in pharmaceutical and clinical research). (Right) Selection of sensor chips with different biospecific features over the gold surface (© 2020 Cytiva—Reproduced by permission of the owner)

Compared to other techniques, like the heterogeneous immunoassays, which only provide end-point results, SPR allows understanding the entire interaction between molecules, providing answers to *four main questions*:

- *How much?* SPR determines the concentration of active molecules in the sample. Active Concentration Analysis (ACA) + General Quantitative Analysis (GQA).

- *How strong?* Affinity determinations provide details on how the biological process occurs and is controlled.
- *How fast?* Kinetic measurements give information on the rates at which interacting molecules bind (associate) and break apart (dissociate).
- *How specific?* Examine the ability of a molecule to bind to another molecule to the exclusion of others. Of particular interest in structure-function studies.

6.5.1 The Langmuir Isotherm

The Langmuir isotherm describes a monolayer coverage, under the assumption that the adsorbed species do not interact with each other. Thus, this mathematical tool can be conveniently adopted for the assessment of the equilibrium constants and the kinetic rates of affinity binding.

In the case of a "static" observation, that is at equilibrium, the Langmuir isotherm is expressed as follows:

$$\Gamma_e = \Gamma_m \left(\frac{K_A C}{1 + K_A C} \right) = \Gamma_m \left(\frac{C/K_D}{1 + C/K_D} \right)$$

with Γ_e surface coverage at equilibrium, Γ_m: full monolayer coverage (saturation), K_A: equilibrium association constant, K_D: equilibrium dissociation constant, C: concentration of analyte (Fig. 6.17).

In the case of a "dynamic" observation, $\Gamma(t)$ describes the kinetics of adsorption and is formulated in this way:

$$\Gamma(t) = \Gamma_e[1 - \exp(-k_{on} C_t)]$$

with $\Gamma(t)$ surface coverage at any given time, k_{on} association rate constant, C_t concentration at any given time.

Once assessed, the equilibrium constants deliver relevant and quantitative answers to fundamental questions related to the specificity and the strength of the affinity binding. This is particularly beneficial in the pharmaceutical research for the characterization and the selection of tailored antibodies (see Table 6.1 for more details).

6.5.2 SPR Parameters and Kinetic Analysis

In addition to the determination of the equilibrium as seen above, the analysis of the SPR sensogram delivers the kinetic rate constants k_{on} and k_{off}, thus answering the question: How FAST is the binding? Look at the *rate constants*:

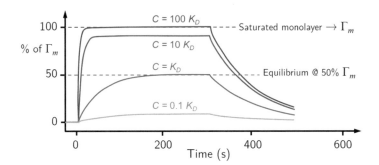

Fig. 6.17 Equilibrium Information obtained by running multiple measurements with different analyte concentrations: If $\Gamma_e = 0.5 \, \Gamma_m$ then $K_D = C$ and $K_A = 1/K_D$

Table 6.1 Properties assessed by means of equilibrium experiments (i.e., without kinetics)

Specificity	• Do two molecules interact with each other? Yes/No answers • Different analytes are tested with the same ligand • Quantitative measurements, test different analyte concentrations to determine the concentration dependency of the response
Concentration assays	• Concentration based on biological activity • All concentration assays require a calibration curve • Concentrations of unknowns samples are calculated from this • 4–7 concentrations in duplicate • Moderate to high densities on sensor chip • Choice of formats: Direct binding/competition/inhibition
Affinity constants	• How STRONG is the binding at equilibrium? • Quantify K_A and K_D • Rank antibodies • Find best Ab pairs

$k_{on} = k_a = k_1$ (association, recognition)[1]
$k_{off} = k_d = k_{-1}$ (dissociation, stability)

The procedure for extracting the rate constants from sensograms follows the next steps:

• Measure binding curves, that is, the response over time R_t for various analyte concentrations.
• Select a model to describe the interaction.

---

[1] In literature different authors use different denominations for the same entity, therefore the respective equivalences $k_{on} = k_a = k_1$ and $k_{off} = k_d = k_{-1}$ are explicitly given here for clarity.

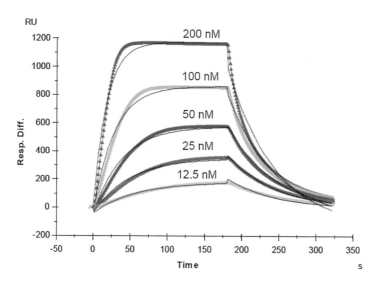

Fig. 6.18 Extracting rate constants from sensograms: Color curves are experimental data and black curves are the fitting curves using 1:1 Langmuir binding model from BIAevaluate software (Source: [12], © by the authors, CC-BY-4.0 [13])

Fit the curve to a rate equation describing the model; for example, 1:1 binding:

$$\frac{dR}{dt} = k_{on}C(R_{max} - R_t) - k_{off}R_t$$

- Obtain values for k_{on}, k_{off}, R_{max}.
- Assess the fit:
  - Overlay experimental plots and residual plots.
  - Estimate quality of statistics, for example, x^2-curve fidelity: Gives a measure of the accuracy of the fitting and is determined over all curves fitted simultaneously (Fig. 6.18). Hereby, the x^2 values should be lower than about 10 for a good fit to sensograms with normal noise levels.
  - Obtain the biological and experimental relevance of the calculated parameters.

The kinetic rate constants show how fast the ligands diffuse and bind to the immobilized receptors. However, the strength of this binding is solely given by the equilibrium constants. In fact, accordingly to the relationship between rate and equilibrium constants in Table 6.2, it is possible to observe the same affinity strength from compounds having rate constants, which differ by orders of magnitude.

As mentioned above, this method is particularly important for the screening of engineered antibodies for therapeutic purposes. Figure 6.19 shows kinetic binding analysis

Table 6.2 Equilibrium and kinetic constants are related to each other

Parameter	Association rate constant k_{on}	Dissociation rate constant k_{off}	Equilibrium dissociation constant K_D	Equilibrium association constant K_A
Definition	$[A] + [B] \xrightarrow{k_{on}} [AB]$	$[AB] \xrightarrow{k_{off}} [A] + [B]$	$\frac{[A][B]}{[AB]} = \frac{k_{off}}{k_{on}}$	$\frac{[AB]}{[A][B]} = \frac{k_{on}}{k_{off}}$
Units	$[M^{-1}s^{-1}]$	$[s^{-1}]$	$[M]$	$[M^{-1}]$
Description	Rate of complex formation, i.e., the number of AB formed per second in a 1 molar solution of A and B	Stability of a complex, i.e., the fraction of complexes that decays per second	Dissociation tendency High $K_D =$ low affinity	Association tendency High $K_A =$ high affinity
Typical range	1×10^{-3}–1×10^{7}	1×10^{-1}–5×10^{-6}	1×10^{-5}– 1×10^{-12}	1×10^{5}– 1×10^{12}

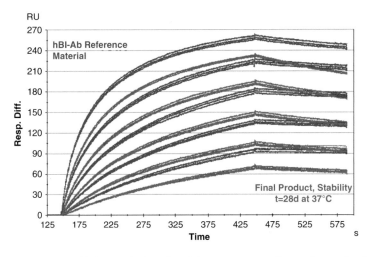

Fig. 6.19 Kinetic and affinity analysis using Biacore system: Comparison of reference Ab (blue) vs. packaged product after 28 day of thermal stress (red). The protocol was run in triplicate for assessing the reproducibility. The rate constants k_{on} and k_{off} of the stressed product indicate 34% slower association and 3.1 times faster dissociation than the reference. This is an important information regarding the shelf life of the product (© Boehringer Ingelheim—Reproduced by permission of the owner)

data of a pharmaceutical reference Ab in comparison with the same Ab stored for 28 days under thermal stress at 37 °C.

Since different antibodies may bind to different specific epitope features of the same antigen (multi-site binding), it is further possible to investigate the behavior of a variety of antibodies in sequential runs for determining the respective binding strength, even in

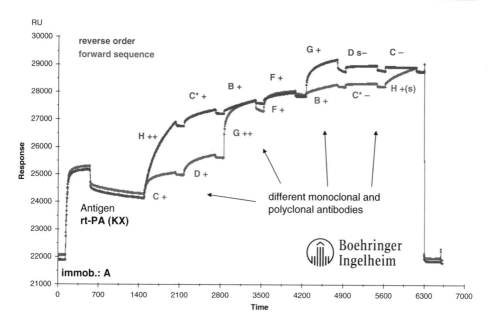

Fig. 6.20 Epitope mapping by MSBA. The antigen rt-PA is captured by the immobilized primary antibody A. Seven antibodies with different affinities bind sequentially to the antigen. Different responses in the forward and reverse sequence indicate overlap of competing binding regions (© Boehringer Ingelheim—Reproduced by permission of the owner)

competition with each other. In fact, some antibodies may partially share the same epitope feature, leading to some degree of overlap.

By adopting a forward and reverse sequence in two separate runs of the so-called *Multi-Site Binding Analysis* (MSBA), the collected information is even more exhaustive. An example of this method is the behavioral study of the therapeutic protein rt-PA [2] (recombinant tissue–type Plasminogen Activator) in the presence of a variety of antibodies (see Fig. 6.20).

The observed affinity bindings are responsible for the inactivation of the drug (not shown), which has a half-life of only 4–9 min, and the activation of the complement system which may lead to severe side effects like the reocclusion of the blood vessels and damages to platelets and vascular endothelial cells.

Question 3

Can you explain the meaning of the four main questions to which the SPR method can provide the answers?

---

[2] Actilyse® (rt-PA, alteplase): Thrombolytic treatment in case of acute myocardial infarction, acute massive pulmonary embolism, and acute ischemic stroke.

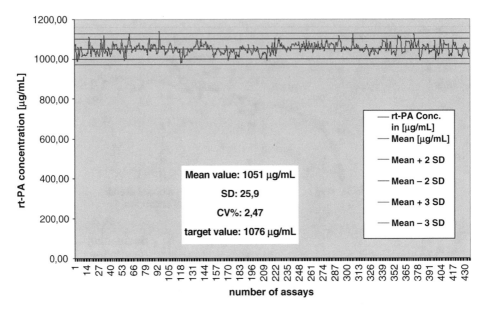

Fig. 6.21 Trending chart over a 440 days timespan obtained by SST with a reference concentration of rt-PA. The nearly constant mean and the standard deviation of ~2.5% indicate a high reliability of the capturing matrix (© Boehringer Ingelheim—Reproduced by permission of the owner)

Control of correct assay procedure and the control of matrix stability can be assessed by running a System Suitability Test (SST), that is, measuring a control sample for every individual assay. Furthermore, by collecting SST data and assay representative data of all individual assays, it is possible to determine whether there is a long-term trend, that is, whether the test method shows trends over a long time. However, these trends are more likely to indicate that the critical reagents used (immobilized receptor, reference material, and control sample) experience a storage-related change in activity (Fig. 6.21).

6.6   Gravimetric Biosensors

Gravimetric biosensors are another family of devices and systems for detecting label-free analytes. They work by sensing a change in mass. This can be obtained with static sensors by measuring the deflection/force or with dynamic sensors by detecting the shift of the resonance frequency.

6.6.1   Cantilever Biosensors

Cantilever sensors are microfabricated beams clamped at one end. At the other end, the surface is functionalized with bioreceptors, in most cases for affinity binding. When a sample is dispensed and recognized analyte molecules are bound, the mass increases. The detection of the extra mass can be in deflection mode, that is, measure the quasi-static bending, or in resonance mode, with increasing mass a decrease of the resonant frequency is detected [14]. In most cases, the transducer consists in a piezoresistive detector, built-in close to the clamp, which is the position with maximum stress. It is convenient to adopt a Wheatstone bridge configuration for the detector [15], because this design provides compensation of thermal drift and a higher sensitivity. The lever effect provided by the beam, that is, load at the tip and read-out at the clamp, results in a mechanical amplification of several orders of magnitude, which makes these structures sensitive to very small masses, typically from the nanogram down to the picogram range or even below. Other read-out methods include laser-beam deflection [16] or a piezoelectric detector [17]. Cantilevers are suitable for arraying up to a few tens of units, thus enabling simultaneous multi-analyte detection [18, 19] (Figs. 6.22 and 6.23).

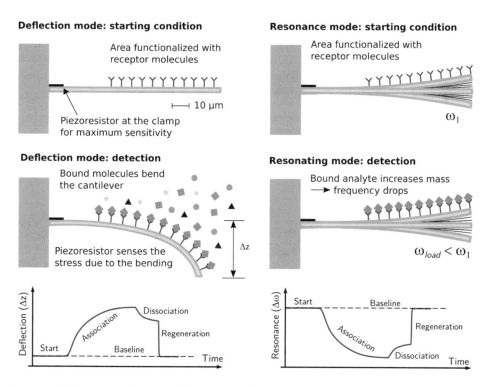

Fig. 6.22 Operating modes of cantilever sensors: Deflection mode (left) and resonant mode (right). Functionalized cantilever working in deflection mode, without bending (left) and bent due to additional mass of bound molecules (right)

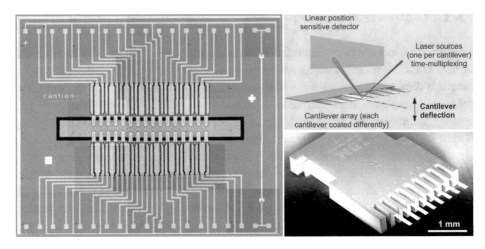

Fig. 6.23 Left) Bilateral array of 32 piezoresistive cantilevers with a microfluidic channel (Reprinted from [19], © 2009, with permission from Elsevier). (Right) Single-side array of eight cantilevers designed for detection via laser-beam reflection. The top mirror layer is out of gold, which allows at the same time immobilization of thiolated receptors via sulfur-bond on gold (Reprinted from [20], © 2005, with permission from Elsevier)

Fig. 6.24 Stress profile in the cantilever as reaction to the applied force due to mass-loading

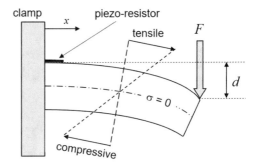

For analyzing the deflection mode, we now consider a cantilever of length l, thickness t, and width w clamped at one end. Usually a piezo-resistor is placed on the upper side of the cantilever. The compressive stress at the lower side is equal and opposite of the tensile stress at the upper side (Fig. 6.24).

Here the stress σ at any point x along the cantilever can be calculated with the following expression:

$$\sigma(x) = 6F(l-x)/(t^2 w)$$

The stress is maximum at the clamped end $(x = 0)$, therefore:

$$\sigma_{max} = 6Fl/\left(t^2 w\right)$$

Given π, the piezoresistive coefficient of the piezoresistor, under the above stress the piezoresistor changes its value R as follows:

$$\frac{\Delta R}{R} = \sigma_{max}\pi$$

Finally, the deflection d observed at the free end, where the force F is applied, can be calculated as follows (here E is the Young's modulus of the beam):

$$d = 4Fl^3/\left(Ewt^3\right)$$

For analyzing the operation of a cantilever sensor in resonance mode, we consider again a beam of length l, thickness t and width w clamped at one end. The cantilever material has a density ρ and Young's modulus E, thus we can define:

The moment of inertia of the beam cross-section: $\qquad I = \frac{wt^3}{12}$

The stiffness of the beam (elastic coefficient): $\qquad k = \frac{3EI}{l^3}$

The first natural resonant frequency: $\qquad \omega_1 = \frac{3.516}{l^2}\sqrt{\frac{EI}{\rho wt}}$

Let us consider now an equivalent massless cantilever with a discrete effective mass m_{eff} concentrated at the free end. Under this assumption we obtain:

$$\omega_1 = \sqrt{\frac{k}{m_{eff}}}$$

If the cantilever is loaded now at the free end with a sample of mass Δm, the resonant frequency becomes:

$$\omega_{load} = \sqrt{\frac{k}{m_{eff} + \Delta m}}$$

For example, the cantilevers reported in [21] change their resonating frequency by ~1 Hz/pg.

As seen above, cantilevers are suitable for array designs for detecting multiple analytes simultaneously. For their functionalization and the dispensing of analytes different techniques are available, as shown in Fig. 6.25.

The shift of resonant frequency due to absorption of molecules or a chemical reaction can be used for detection in a variety of biosensors. Beside the cantilevers seen above, such

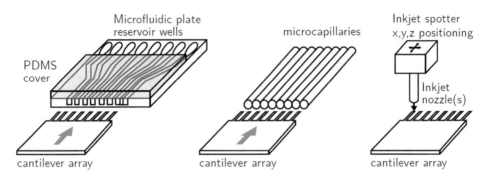

Fig. 6.25 Individual functionalization for multiple simultaneous analysis: Insertion into microfluidic channels (left), insertion into microcapillaries (center), and spotting with an inkjet head (right). For bilateral arrays as shown in Fig. 6.23, only the inkjet spotter is appropriate

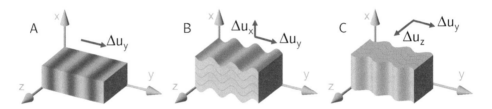

Fig. 6.26 Profiles of particle displacement (Δu) for (**a**) longitudinal (pressure waves) and (**b**) and (**c**) for transverse waves (also called *shear waves*)

resonating sensors make use of piezoelectric crystals, which resonate producing either bulk acoustic waves (BAW), surface acoustic waves (SAW), or even plate acoustic waves (PAW), depending on the material properties and design parameters of the specific device.

These sensors can be operated in a gaseous or liquid environment. In the latter case, a shear mode wave is necessary. In the Fig. 6.26, *bulk acoustic waves* are shown (i.e., constant wave amplitude along the material). The wave propagation is in the Y direction for all three cases.

6.7   Bulk Acoustic Wave (BAW) Sensor

Now let us look at a BAW-based sensor. For this purpose, a quartz crystal with AT-cut [3] [22] is structured with one thin-film electrode at each side for excitation/transduction. For biosensing purposes, this metal layers consists of gold, due to its chemical stability in

---

[3] AT cut: angle of cut of the crystal plate, which yields ~zero temperature coefficient at room temperature.

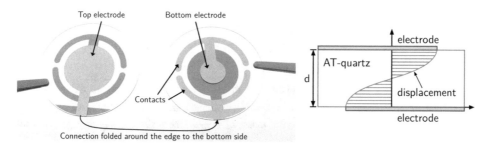

Fig. 6.27 (Left) Quartz crystals structured as BAW sensors. (Right) Shear resonance mode

water-based solutions. The electrode at the top side is larger, in order to provide an appropriate surface for receptor immobilization, typically done with thiolated linkers via sulfur-bond on gold. The crystal resonates in shear mode; this is energetically very convenient, in fact in this way only sliding friction influences the resonance. On the contrary, in case of pressure waves, the whole water mass over the crystal would need to be displaced, which would require a much larger dispense of energy (Fig. 6.27).

For the resonant frequency f_0 we have:

$$f_0 = n\frac{v_b}{2d}$$

v_b is the shear wave velocity (3200 m/s), d is the crystal thickness, and $n = 1, 2, 3, \ldots$ is the fundamental frequency or the respective harmonics.

According to Sauerbrey [23], the mass sensitivity S is given by the following:

$$S = \frac{\Delta f}{\Delta m} = k_1 \frac{f_0^2}{A}$$

where $k_1 = -2.3 \times 10^{-7}$ m^{2}kg^{-1}s and A is the effective area of the crystal (electrodes overlapping).

For a usual QCM crystal with $f_0 = 10$ MHz, $A = 1$ cm^2, and $d = \lambda/2 = 160$ μm, we have a theoretical sensitivity of $S = 230$ Hz/μg. The typical LOD of a QCM apparatus is ~10 pg/mm^2.

To analyze the behavior of the quartz resonator, we make use of the Butterworth-Van Dyke's equivalent circuit as shown below (Fig. 6.28):

For such an oscillator, we now consider the magnitude and phase of the impedance, measured with one surface of the crystal in contact with water, as displayed in Fig. 6.29:

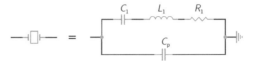

Fig. 6.28 Equivalent circuit for an unperturbed quartz crystal resonator. Where C_p is the static capacitance of electrodes: $C_p = \varepsilon_{22} \frac{A}{d}$, C_1 represents a dynamic capacity for the vibration: $C_1 = \frac{8K2C_p}{n2\pi2}$, the inductance L_1 corresponds to the crystal mass: $L_1 = \frac{1}{\omega_0^2 C_1}$, the resistance R_1 corresponds to the internal friction: $R_1 = \frac{\eta_q \omega2}{\omega_0^2 C_1 C_{66}}$. The constants used in the above formulas are as follows: ε_{22} permittivity of quartz (3.982×10^{-11} F/m), K_2 electromechanical coupling coefficient (7.74×10^{-3}), η_q viscosity of quartz (3.46×10^{-4} Pas), n order of harmonics, $\omega_0 = 2\pi f_0$, C_{66} elastic constant of quartz (2.947×10^{-10} N/m^2).

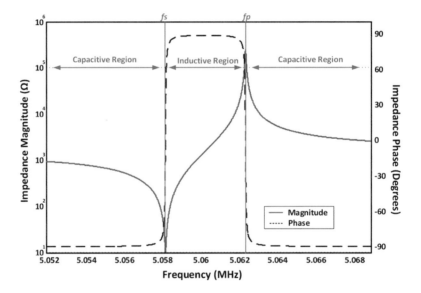

Fig. 6.29 Impedance plot (magnitude and phase) of a BAW quartz resonator. Serial resonance is at 5.058 MHz, parallel resonance at 5.062 MHz. The serial resonance has a very low impedance and is therefore preferred because of lower noise and higher stability (Reprinted from [24], © 2017 by the authors. Licensee MDPI, CC-BY 4.0 [13])

6.8   Quartz Crystal Microbalance (QCM)

Adding analyte results in a modified equivalent circuit. In addition to the existing circuit components seen above, we have now to consider L_2 and R_2 for liquid loading and L_3 for mass loading, that is, the molecular binding layer (Fig. 6.30):

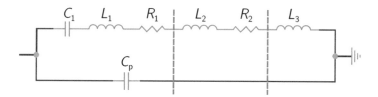

Fig. 6.30 Equivalent circuit of a loaded quartz resonator

L_2 is the mass effect of liquid loading:

$$L_2 = \frac{\omega_0 L_1}{n\pi} \sqrt{\frac{2\rho_l \eta_l}{\omega C_{66}\rho_q}}$$

R_2 is the friction effect due to the liquid viscosity:

$$R_2 = \frac{\omega_0 L_1}{n\pi} \sqrt{\frac{2\omega\rho_l \eta_l}{C_{66}\rho_q}}$$

L_3 is the mass load due to molecular binding:

$$L_3 = \frac{2\omega_0 L_1 \rho_s d_s}{n\pi \sqrt{C_{66}\rho_q}}$$

with ρ_l density of liquid, η_l viscosity of liquid, ρ_s density of mass layer, d_s thickness of the mass layer, ρ_q density of quartz (2.65×10^3 kg/m^3).

It is important to understand that loading the resonator with water and additional molecular binding leads to the necessity of an electronic driving unit able to keep the crystal in resonance mode under large variations of mass and viscosity. To this purpose, appropriate designs must include, among others, adaptive feedback and automatic gain and phase control circuits. Comprehensive discussions of these concepts and specifications are provided in [24, 25].

At this point, we have all the elements for understanding how a BAW resonator can be used for measuring minute changes of mass due to molecular binding, as in affinity biosensors. Such an implementation is named Quartz Crystal Microbalance (QCM) and was introduced by G. Sauerbrey in 1959 [23]. The QCM-sensogram is conceptually identical to the one of resonating cantilevers; therefore, also QCM is a real-time method, able to deliver information on the kinetics of a reaction, like TIRF and SPR.

According to the Sauerbrey's equation, the QCM sensitivity is proportional to the square of the resonance frequency f_0, which in turn is inversely proportional to the crystal thickness d. However, there are limitations to the improvement of sensitivity achievable by

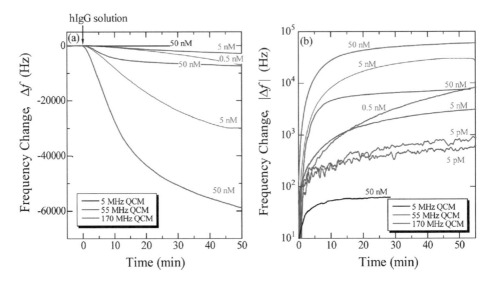

Fig. 6.31 Linear (left) and logarithmic (right) frequency change of three electrodeless QCM crystals, of 30, 50, and 300 μm, resonating at 170, 55 and 5 MHz respectively, vs. time for different concentrations of IgG solution (Reprinted from [26], © 2009, with permission of the American Chemical Society)

thinning the crystal, not only because of increasing fragility, but also because the metal electrodes represent themselves a load whose impact becomes relevant with thin quartz plates. An alternative approach which overcomes such problems is the electrodeless, inductively coupled QCM [26] for which resonating frequencies of 170 MHz have been reported (Fig. 6.31).

Question 4

Draw the equivalent circuits for a BAW sensor with and without absorbed molecules and explain the additional terms that add up during absorption.

6.8.1   QCM-D Principle

Adsorbed analytes can spoil the resonance quality factor Q, due to frictional energy dissipation (Fig. 6.32). Therefore, the driving circuit of the QCM includes an automatic gain control (ACG) for continuously compensating the energy loss and keeping the crystal in resonance. A simple method for assessing the viscosity of the analyte under test consists in monitoring the working gain level.

A more precise method to study the viscoelastic properties of a sample consists in the transient analysis of the energy dissipation in the vibration system. The dissipation is defined as follows:

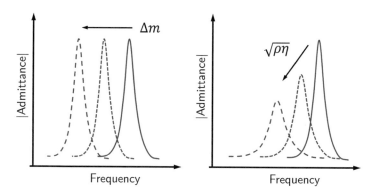

Fig. 6.32 Shift of resonance frequency due to mass change (left) and energy loss (amplitude decrease) depending on viscosity η and density ρ (right)

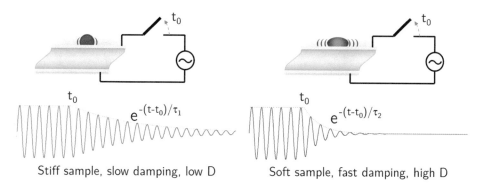

Fig. 6.33 Transient behavior after disconnecting the excitation source at time t_0: A stiff sample has low dissipation which results in a long decay-time (left), while a soft sample dissipates the vibrational energy faster (right)

$$D = \frac{E_{\text{lost}}}{2\pi E_{\text{stored}}}$$

where E_{lost} is the energy loss during an oscillation cycle and E_{stored} is the total energy stored in the oscillator. Therefore, in a QCM the frequency change (Δf) depends on the mass while the change in dissipation (ΔD) depends on the viscoelastic properties. The *QCM-D* method [27] is based on the transient analysis after disconnecting the crystal from the driving circuit (Fig. 6.33). Out of the detected decay time τ, the dissipation is calculated with the following equation [28]:

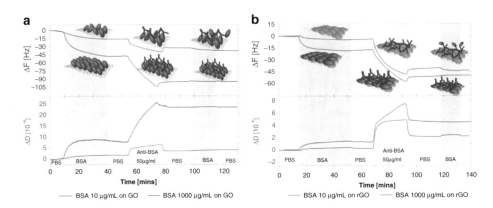

Fig. 6.34 QCM-D monitoring of BSA and anti-BSA interaction on (**a**) GO and (**b**) rGO. The protein adsorption on GO can maintain binding functionality to its antibody. The denaturation of protein was found on highly hydrophobic rGO, but an additional layer of protein can be created on top of the denatured layer without denaturation (Reprinted from [29], © 2020 by the Authors, published by Elsevier Ltd. under CC-BY-4.0 [13])

$$D = \frac{1}{\pi f \tau}$$

Compared to previously discussed methods, the QCM-D technology allows, in addition to the assessment of the equilibrium and kinetic constants, the quantitative analysis of thickness, shear elastic modulus, and viscosity of the adsorbed layer.

As an example, Fig. 6.34 shows the time courses for change of frequency and dissipation observed during the adsorption of bovine serum albumin (BSA) on hydrophilic graphene oxide (GO) and hydrophobic reduced graphene oxide (rGO).

Another important feature of the QCM method is the possibility to use the top electrode of the quartz resonator for electrochemical protocols. As examples, this is useful for growing a layer of linker molecules or a polymeric film for anchoring bioreceptors in a next step, with the possibility of monitoring very precisely and in real time via QCM the mass and the viscoelastic properties of such layers [30–32]. This method is called *EQCM* (Fig. 6.35).

In summary, the QCM method can quantitatively analyze:

- Surface adsorption and desorption
    - Biomolecules (protein, vitamin, antibodies, DNA, etc.)
    - Polymers, polyelectrolytes
    - Cells
- Surface reactions (EQCM)
    - Conformational change (protein, DNA, polymer, cells).
    - Growth of monolayers (linkers)

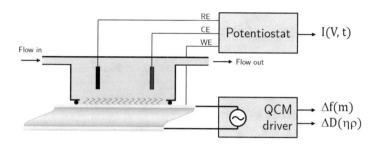

Fig. 6.35 The EQCM technology allows combinations of electrochemical and gravimetric protocols. The gold electrode at the top side of the BAW resonator acts at the same time as working electrode (WE) of the electrochemical setup

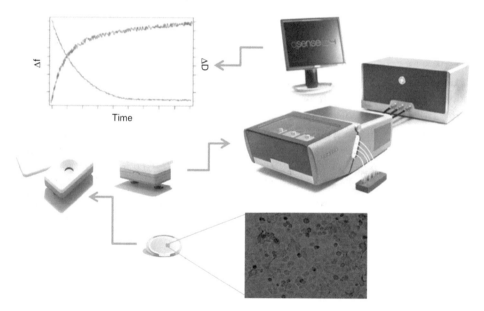

Fig. 6.36 Q-Sense 4 of Biolin Scientific Group (Reprinted from [38], © 2015, with permission from Springer Science+Business Media)

- Cross-connections (protein, polymer, etc.)
- Hydration (polymer)
- Bulk characterization (QCM-D)
  - Viscoelastic properties of liquids (protein solution, etc.)

There are several commercial companies offering QCM systems for different application scenarios and budgets. Among others we find: Biolin Scientific with the Q-Sense product line [33] (Fig. 6.36), Nihon Dempa Kogyo with the NAPiCOS line [34], 3T-Analytik with the qCell lines [35], Gamry Instruments with the QCM-I and EQCM-I lines [36] and Stanford Research Systems with the QCM200 [37].

6.9   Surface Acoustic Waves

In surface acoustic waves (SAW) devices, the energy of the mechanical wave is concentrated on the surface of the material. The material thickness is much larger than the wavelength (Fig. 6.37).

6.9.1   SAW Sensors

A SAW sensor consists of two electrode groups having an interdigitated pattern, fabricated in thin-film technology on suitable materials like polished quartz crystal, bulk lithium niobate ($LiNbO_3$), and lithium tantalate ($LiTaO_3$) or piezoelectric thin films of aluminum nitride (AlN). The RF voltage applied at one *Inter Digitated Transducer* (IDT) generates a shear acoustic wave, which propagates along the surface and can be measured as a voltage at the other IDT. Reflector structures placed beyond the IDT's help in confining the energy in the sensitive region (Fig. 6.38). A mass change at the surface in the detection region between the IDT's causes a frequency shift.

Typical SAW devices have propagation length L and substrate thickness d of a few tenths of mm. The pitch of the IDT λ is a design parameter, ranging from few µm to tens of µm, which defines the resonance frequency f_0 of the device:

$$f_0 = \frac{v_0}{\lambda}$$

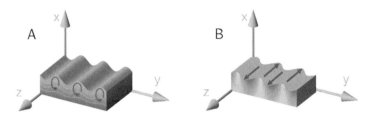

Fig. 6.37 Surface acoustic waves: (**a**) Rayleigh (after the British physicist Lord J. W. S. Rayleigh) waves are a combination of pressure and vertical shear waves. The particles follow an elliptical orbit, approximately circular at the surface. (**b**) Love (after the British mathematician A. E. H. Love) waves, on the other hand, are horizontal shear waves. In both types of waves, the energy profile decreases exponentially with depth

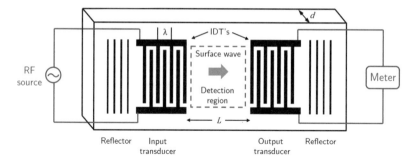

Fig. 6.38 Scheme of a SAW sensor

where v_0 is the velocity of the acoustic wave for the given substrate material in unloaded state. The presence and the dynamic change of a biosensitive layer in the detection region alter the propagation velocity of the acoustic waves, which results in a detectable frequency shift.

The mass sensitivity S can be calculated again with the Sauerbrey equation, whereas the material constant k is now specific for the SAW propagation; for a quartz plate it is $k = -1.35 \times 10^{-7}$ m^{2}s/kg. For example, with a sensor fabricated on a quartz crystal with an IDT design resonating at 200 MHz, we obtain a mass sensitivity $S = -5.4$ Hz pg^{-1} mm^{-2} and an LOD of ~3 pg/mm^2, assuming a signal-to-noise ratio of 3 and a bandwidth of 10 Hz for the signal acquisition.

When designing a SAW sensor it is important to analyze the electromechanical coupling k^2 at the IDTs. This important parameter depends on the properties of the adopted piezoelectric material and can be calculated with the following formula [40]:

$$k^2 = \frac{e_{31}^2}{c_{11}\varepsilon_{33}}$$

where e_{31}^2, c_{11}, and ε_{33} are respectively the components of the material property tensors for piezoelectric constant, mechanical stiffness, and permittivity involved in the coupling process.

From an experimental point of view, k^2 can be determined on the basis of design parameters and measurement data as follows:

$$k^2 = \frac{\pi}{4N}\left(\frac{G}{B}\right)_{f=f_0}$$

where N is the number of the finger pairs in the IDT, G is the conductance, and B the susceptance, that is, real and imaginary part of the device admittance at the central frequency, respectively [40].

Table 6.3 Properties of some of the most common materials for SAW devices [39]

Material	Relative permittivity	Temperature coefficient of frequency (ppm/°C)	Velocity (m/s)
Quartz	3.8	~0	3159
LiTaO$_3$ (X-112° Y)	43	18	3300
LiTaO$_3$ (36° Y-X)		32	4160
LiNbO$_3$ (128° Y-X)	85.2	75	3979
LiNbO$_3$ (64° Y-X)		80	4742
LiNbO$_3$ (Y-Z)		94	3488
AlN	8.5	19	5700

As shown in Table 6.3, quartz has nearly no thermal drift of frequency, but its low permittivity makes it unsuitable for operation in contact with a water-based medium due to the much larger relative permittivity of water (~80), which spoils the electromechanical coupling. Lithium tantalates have intermediate thermal drift, depending on the crystal cut, and a higher dielectric constant. Lithium niobates have the highest dielectric constant, which makes them the best choice for operation in contact with water, but poor thermal drift, and aluminum nitride has a relatively low thermal drift and the highest SAW velocity.

Due to the above properties, SAW devices are most commonly adopted for analytical purposes in gas and vapor phases, for example, for the detection of volatile biomarkers in breath for assessing cancer risk [41] or the identification of herbal aromas for controlling the quality of pharmaceutical plants [42], or even to assess in air the presence of pollutants [43] or chemical warfare agents [44]. Accordingly, commercial systems called **electronic noses** are available on the market, like the zNose line by Electronic Sensor Technology [45], the HAZMATCAD handheld by Enmet [46], or the SAGAS of the Karlsruhe Institute of Technology [47]. In the last years, the efforts in the development of SAW devices for liquid phase sensing increased significantly due to their potential of being more sensitive than BAW sensors; however the related technologies have not yet become established and therefore their use is still confined in the research area.

Question 6

How does a SAW sensor work? How is its mass sensitivity defined?

Answer 1

The Mach–Zehnder interferometer exploits the change of refractive index (RI) in the sensing region due to the presence of a biolayer, which absorbs energy from the evanescent wave. This variation of RI alters the velocity of light in the sensing arm, which produces an interference at the Y-junction where the two half beams recombine. For proper function, a coherent laser light source is necessary.

The light intensity at the output is given by $I = (I_0/2)[1 + \cos(\Delta\varphi)]$, where the phase shift is given by $\Delta\varphi = 2\pi\Delta n\, L/\lambda$. Therefore, the phase shift is the measured physical quantity.

The amount of bound analyte is qualitatively proportional, although non-linearly, to the phase shift. For a quantitative response, a calibration is required.

Answer 2

The dispersion relation for air has no intersection with the dispersion relation of surface plasmon in the metal layer. In order to obtain an intersection, that is, matching the wavevector of the excitation with the wavevector of the plasmon, the speed of light must be reduced, so the dispersion relation is conveniently tilted. This can be easily reached with the help of a medium with a higher refractive index than the one of air. Optical grade glass is a common choice ($RI \sim 1.5$), but also other optical materials with suitable RI will work.

Answer 3

The four main questions (How much? How strong? How fast? How specific?) which arise when analyzing an immunoassay are very relevant for a comprehensive characterization of a receptor–ligand interaction. In particular, they play a fundamental role in the development of affinity-based medications like therapeutic antibodies.

Answer 4

In the equivalent circuit, we find the three typical components of a resonator, namely the capacitor that represents the elasticity, the inductor, equivalent to the inertial mass and the resistor that represents the internal friction in the crystal. In parallel, we need to add the extra capacity due to the electrode plates used for excitation/sensing. When the BAW resonator is interfaced to a liquid phase, two additional components come into play, namely the mass effect of the liquid represented by a second inductor and the friction effect symbolized with a second resistor. Finally, when receptors and ligands form the biolayer, we represent their mass with a third inductor.

Answer 5

QCM-D can assess the viscoelastic properties of an analyte. It works based on the transient analysis after disconnecting the driving circuit from the resonator and detecting the decay-time of the resonance. Out of the measured time constant, it is possible to calculate the dissipation with the formula $D = 1/(\pi f \tau)$.

Answer 6

A SAW sensor is another gravimetric device, fabricated on suitable piezoelectric materials. Here the acoustic waves propagate predominantly on the surface and their amplitude decays exponentially in a shallow depth. The frequency is defined by the design parameter λ, that is, the pitch of the interdigitated transducers (IDT). There are two IDTs—one for excitation, the other for readout. The sensing area is between the two. Loading the sensor with a mass changes the propagation velocity of the acoustic waves, which results in a detectable frequency shift. The sensitivity is defined the same way of QCM, that is, $S = \Delta f/\Delta m$ and is calculated with the Sauerbrey equation, in which the constant k is given by the SAW-related behavior of the material in use.

Take-Home Messages
- Label-free detection is possible whenever the analyte can alter a measurable property of a physical phenomenon.
- In the Mach–Zehnder interferometer, the measurable physical property is the phase shift due to change of the refractive index when biomolecules bind.
- In the SPR method, a change of the RI alters the matching conditions of light wavevector and plasmon resonance wavevector. Depending on the adopted setup, this effect results either in a shift of the angle or in a shift of the wavelength at which the maximum light absorption is observed.
- Gravimetric sensing methods detect changes of bound mass mechanically. Cantilevers, for instance, bend their beam proportional to the load when operated in static mode or lower their resonance frequency when used in dynamic mode.
- Also, BAW and SAW devices work in resonating mode. The quartz crystal microbalance is a gravimetric method with a BAW resonator as core device.
- QCM can be extended with other capabilities, like the assessment of viscoelastic properties (QCM-D) and/or combined with a potentiostat for concurrent electrochemical protocols (EQCM).
- SAW devices have the potential to be more sensitive than their BAW relatives, but the technology for biosensing in the water phase is not yet fully established.

References

1. Kozma P, Kehl F, Ehrentreich-Förster E, Stamm C, Bier FF. Integrated planar optical waveguide interferometer biosensors: a comparative review. Biosens Bioelectron. 2014;58:287–307.
2. Prieto F, Sepúlveda B, Calle A, Llobera A, Dominguez C, Lechuga LM. Integrated Mach–Zehnder interferometer based on ARROW structures for biosensor applications. Sens Actuators B. 2003;92(1):151–8.
3. Homola J. Electromagnetic theory of surface plasmons. In: Surface plasmon resonance based sensors. New York: Springer; 2006.
4. Bolduc OR, Live LS, Masson J-F. High-resolution surface plasmon resonance sensors based on a dove prism. Talanta. 2009;77(5):1680–7.

5. Dodel N, Laifi A, Eberhardt D, Keil S, Sturm L, Klumpp A, et al. A CMOS-based electrochemical DNA hybridization microarray for diagnostic applications with 109 test sites. In: 2019 IEEE biomedical circuits and systems conference (BioCAS); 2019 17–19 Oct 2019.
6. Drummond TG, Hill MG, Barton JK. Electrochemical DNA sensors. Nat Biotechnol. 2003;21 (10):1192–9.
7. Schienle M, Paulus C, Frey A, Hofmann F, Holzapfl B, Schindler-Bauer P, et al. A fully electronic DNA sensor with 128 positions and in-pixel A/D conversion. IEEE J Solid State Circuits. 2004;39(12):2438–45.
8. Abe T, Esashi M. One-chip multichannel quartz crystal microbalance (QCM) fabricated by deep RIE. Sensors Actuators A Phys. 2000;82(1):139–43.
9. Furst A, Landefeld S, Hill MG, Barton JK. Electrochemical patterning and detection of DNA arrays on a two-electrode platform. J Am Chem Soc. 2013;135(51):19099–102.
10. Tuantranont A, Wisitsora-at A, Sritongkham P, Jaruwongrungsee K. A review of monolithic multichannel quartz crystal microbalance: a review. Anal Chim Acta. 2011;687(2):114–28.
11. Sjoelander S, Urbaniczky C. Integrated fluid handling system for biomolecular interaction analysis. Anal Chem. 1991;63(20):2338–45.
12. Zhao J, Kong Y, Zhang F, Linhardt RJ. Impact of temperature on heparin and protein interactions. Biochem Physiol. 2018;7:2.
13. The Creative Commons copyright licenses. https://creativecommons.org/licenses/.
14. Goeders KM, Colton JS, Bottomley LA. Microcantilevers: sensing chemical interactions via mechanical motion. Chem Rev. 2008;108(2):522–42.
15. Mukhopadhyay R, Lorentzen M, Kjems J, Besenbacher F. Nanomechanical sensing of DNA sequences using piezoresistive cantilevers. Langmuir. 2005;21(18):8400–8.
16. Tian F, Hansen KM, Ferrell TL, Thundat T, Hansen DC. Dynamic microcantilever sensors for discerning biomolecular interactions. Anal Chem. 2005;77(6):1601–6.
17. Yoo YK, Chae M-S, Kang JY, Kim TS, Hwang KS, Lee JH. Multifunctionalized cantilever systems for electronic nose applications. Anal Chem. 2012;84(19):8240–5.
18. Gruber K, Horlacher T, Castelli R, Mader A, Seeberger PH, Hermann BA. Cantilever array sensors detect specific carbohydrate–protein interactions with Picomolar sensitivity. ACS Nano. 2011;5(5):3670–8.
19. Boisen A, Thundat T. Design & fabrication of cantilever array biosensors. Mater Today. 2009;12 (9):32–8.
20. Lang HP, Hegner M, Gerber C. Cantilever array sensors. Mater Today. 2005;8(4):30–6.
21. Battiston FM, Ramseyer JP, Lang HP, Baller MK, Gerber C, Gimzewski JK, et al. A chemical sensor based on a microfabricated cantilever array with simultaneous resonance-frequency and bending readout. Sens Actuators B. 2001;77(1):122–31.
22. Janshoff A, Galla H-J, Steinem C. Piezoelectric mass-sensing devices as biosensors—an alternative to optical biosensors? Angew Chem Int Ed. 2000;39(22):4004–32.
23. Sauerbrey G. Verwendung von Schwingquarzen zur Wägung dünner Schichten und zur Mikrowägung. Z Phys. 1959;155(2):206–22.
24. Alassi A, Benammar M, Brett D. Quartz crystal microbalance electronic interfacing systems: a review. Sensors. 2017;17:12.
25. Arnau A. A review of interface electronic systems for AT-cut quartz crystal microbalance applications in liquids. Sensors. 2008;8(1):370–411.
26. Ogi H, Naga H, Fukunishi Y, Hirao M, Nishiyama M. 170-MHz electrodeless quartz crystal microbalance biosensor: capability and limitation of higher frequency measurement. Anal Chem. 2009;81(19):8068–73.

27. Rodahl M, Kasemo B. A simple setup to simultaneously measure the resonant frequency and the absolute dissipation factor of a quartz crystal microbalance. Rev Sci Instrum. 1996;67 (9):3238–41.
28. Voinova MV, Rodahl M, Jonson M, Kasemo B. Viscoelastic acoustic response of layered polymer films at fluid-solid interfaces: continuum mechanics approach. Phys Scr. 1999;59(5):391–6.
29. Hampitak P, Melendrez D, Iliut M, Fresquet M, Parsons N, Spencer B, et al. Protein interactions and conformations on graphene-based materials mapped using a quartz-crystal microbalance with dissipation monitoring (QCM-D). Carbon. 2020;165:317–27.
30. Marx KA, Zhou T, Long D. Electropolymerized films formed from the amphiphilic decyl esters of d- and l-tyrosine compared to l-tyrosine using the electrochemical quartz crystal microbalance. Biomacromolecules. 2005;6(3):1698–706.
31. Wang H, Wang J, Choi D, Tang Z, Wu H, Lin Y. EQCM immunoassay for phosphorylated acetylcholinesterase as a biomarker for organophosphate exposures based on selective zirconia adsorption and enzyme-catalytic precipitation. Biosens Bioelectron. 2009;24(8):2377–83.
32. Zeng H, Yu J, Jiang Y, Zeng X. Complex thiolated mannose/quinone film modified on EQCM/Au electrode for recognizing specific carbohydrate–proteins. Biosens Bioelectron. 2014;55:157–61.
33. Homola J, Vaisocherová H, Dostálek J, Piliarik M. Multi-analyte surface plasmon resonance biosensing. Methods. 2005;37(1):26–36.
34. Gao Z, Agarwal A, Trigg AD, Singh N, Fang C, Tung C-H, et al. Silicon nanowire arrays for label-free detection of DNA. Anal Chem. 2007;79(9):3291–7.
35. Microarray2. https://commons.wikimedia.org/wiki/File:Microarray2.gif.
36. Lietard J, Ameur D, Damha MJ, Somoza MM. High-density RNA microarrays synthesized in situ by photolithography. Angew Chem Int Ed. 2018;57(46):15257–61.
37. Song Y, Panas RM, Hopkins JB. A review of micromirror arrays. Precis Eng. 2018;51:729–61.
38. Chen JY, Garcia MP, Penn LS, Xi J. Use of the quartz crystal microbalance with dissipation monitoring for pharmacological evaluation of cell signaling pathways mediated by epidermal growth factor receptors. In: Fang Y, editor. Label-free biosensor methods in drug discovery. New York, NY: Springer; 2015. p. 253–68.
39. Mujahid A, Dickert FL. Surface acoustic wave (SAW) for chemical sensing applications of recognition layers. Sensors. 2017;17:12.
40. Fu YQ, Luo JK, Nguyen NT, Walton AJ, Flewitt AJ, Zu XT, et al. Advances in piezoelectric thin films for acoustic biosensors, acoustofluidics and lab-on-chip applications. Prog Mater Sci. 2017;89:31–91.
41. Phillips M, Cataneo RN, Cruz-Ramos JA, Huston J, Ornelas O, Pappas N, et al. Prediction of breast cancer risk with volatile biomarkers in breath. Breast Cancer Res Treat. 2018;170(2):343–50.
42. Oh SY. An effective quality control of pharmacologically active volatiles of Houttuynia cordata thunb by fast gas chromatography-surface acoustic wave sensor. Molecules. 2015;20(6):10298–312.
43. Tang Y-L, Li Z-J, Ma J-Y, Su H-Q, Guo Y-J, Wang L, et al. Highly sensitive room-temperature surface acoustic wave (SAW) ammonia sensors based on Co3O4/SiO2 composite films. J Hazard Mater. 2014;280:127–33.
44. Matatagui D, Fernández MJ, Fontecha J, Santos JP, Gràcia I, Cané C, et al. Love-wave sensor array to detect, discriminate and classify chemical warfare agent simulants. Sens Actuators B. 2012;175:173–8.
45. Duggan DJ, Bittner M, Chen Y, Meltzer P, Trent JM. Expression profiling using cDNA microarrays. Nat Genet. 1999;21(1):10–4.
46. Spotfire by TIBCO. https://www.tibco.com/products/tibco-spotfire.
47. Yang YH, Dudoit S, Luu P, Lin DM, Peng V, Ngai J, et al. Normalization for cDNA microarray data: a robust composite method addressing single and multiple slide systematic variation. Nucleic Acids Res. 2002;30(4):15.

Bioreceptor Immobilization

7

Contents

Keywords

Interfacial component · Adsorption, entrapment and covalent binding · Surface chemistry · Self-assembled monolayer · Long-term stability · Regeneration

© The Author(s), under exclusive license to Springer Nature Switzerland AG 2021
A. Pasquarelli, *Biosensors and Biochips*, Learning Materials in Biosciences,
https://doi.org/10.1007/978-3-030-76469-2_7

What You Will Learn in This Chapter

In this chapter, you will learn about the various methods used to immobilize the bioreceptor onto a carrier. You will also learn the prerequisites for each method as well as their benefits and drawbacks.

You will be able to identify the correct substance for the desired immobilization method and for optimizing the bond. In addition, you will be able to make predictions on the long-term stability of bonds and on the regeneration of biosensors.

We will deal now with the interfacial component between the bioreceptor and the physical transducer of the biosensor (see Fig. 1.2). Beside the protective and connective functions, the interfacial component is primarily responsible for the bioreceptor immobilization. A successful immobilization protocol should ideally comply with the following requirements:

- Maintain activity and specificity
- Provide a robust coupling
- Facilitate a good signal transduction
- Have an optimized packing density
- Should be obtained with an easy and reproducible procedure

7.1   Immobilization Methods

The three main groups of methods for the immobilization of bioreceptors (e.g., enzymes, antibodies, DNA, etc.) are adsorption, entrapment, and covalent binding. Each one of these methods has advantages and disadvantages compared to the others; therefore, the choice of the best suitable immobilization strategy depends on the specifications and constraints of the targeted application.

These three approaches are schematically depicted in Fig. 7.1.

In the following example, let us examine potential issues related to the immobilization of an enzyme. The active center of the immobilized enzyme must not be deformed and should be ready to receive the substrate molecule (Fig. 7.2a and b), in order to maintain both the activity and specificity of the enzyme. An enzyme immobilized in such a way that the active center is not accessible (Fig. 7.2c) or deformed (Fig. 7.2d), results in an inactive enzyme.

The steric deformation of the active center of the enzymes can be prevented by using fitting template molecules able to keep the active center in its functional conformation during the coupling process, thereby stabilizing the protein and avoiding the unwanted distortion. Also coupling methods that favorably bind to the native and energetically most stable enzyme shape should be preferred.

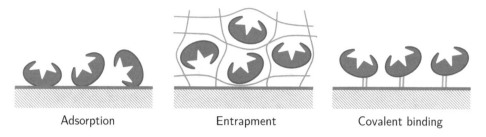

Adsorption   Entrapment   Covalent binding

Fig. 7.1 The three main approaches adopted for immobilization

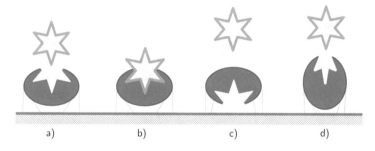

a)   b)   c)   d)

Fig. 7.2 (**a**) Ideally immobilized enzyme: the active site is properly oriented and undistorted, resulting in a catalyst ready to bind and process a substrate molecule, as depicted in (**b**). (**c**) Enzyme immobilized in a way which prevents the substrate from accessing the active site. (**d**) Too tight and rigid bonds may deform the active site thus inactivating the immobilized enzyme

Furthermore, a good strategy for preventing both issues (Fig. 7.2c and d) consists in using an appropriate spacer layer with a convenient permeability for the analyte that creates a diffusion cleft between the enzyme molecules and the carrier structure. This approach dislocate the enzyme away enough from the carrier surface, thus reducing or even eliminating its steric effect.

Question 1
How can be prevented the deformation of active centers during immobilization?

7.2   Adsorption

The *physical adsorption* (also called *physisorption*) is the simplest and cheapest immobilization method. It relies on nonspecific physical interaction (e.g., hydrogen bond, ionic bond, and Van der Waals forces) between the protein and the carrier surface. Since this binding method does not explicitly need reactants, but rather works spontaneously under the normal physiologic chemical conditions, the result is less disruptive to the proteins than chemical adsorption (covalent bond).

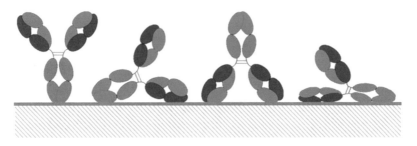

Fig. 7.3 Illustration showing how adsorption may lead to random orientation and denaturation of bioreceptors (antibodies in this case), thus resulting in a decreased total activity

However, the involved bonds are weak, and therefore it is relatively common to observe desorption of the proteins due to changes in temperature, pH, and ionic strength. [1] Other disadvantages are the random orientation (Fig. 7.3) and the nonspecific further adsorption of other proteins or other substances, better called *nonspecific binding* (NSB). This may alter the properties of the immobilized bioreceptor, decrease its activity, and lead to large background signals. Nevertheless, after adsorption immobilization, it is possible to prevent NSB by "blocking," that is, filling the uncovered regions of the carrier surface, for example, by means of a nonfunctional protein like bovine serum albumin (BSA).

Proteins show a strong adsorption on hydrophobic surfaces. This adsorption is often followed by a spontaneous unfolding of the protein structure to increase the hydrophobic portion of the polypeptide chain in contact with the carrier surface. This spontaneous thermodynamic process leads to denaturation with loss of activity in many cases. This effect is less pronounced with antibodies; in fact they have a lower tendency to denaturate than enzymes.

The most favorable surfaces should reproduce the native physiologic aqueous conditions, that is, uncharged environment rich in $-OH$ groups. It is therefore important to "tune" the carrier surface in advance.

Question 2
Which properties characterize the immobilization by means of adsorption?

7.2.1 Surface Modifications

Several carrier material are available for fabricating biosensors. Their native surfaces properties may need a chemical modification depending on the selected immobilization

---

[1] The ionic strength I is a function of the concentrations c_i of *all* ions (of charge z_i) present in a solution: $I = \frac{1}{2}\sum_{i=1}^{n} c_i z_i^2$.

Table 7.1 Carrier materials, their physicochemical properties, and technological applications

Material	Technological use	Physicochemical properties	Modifications
Gold	Compatible with many semiconductor technologies	Native hydrophilic. Tendency to hydrophobicity with time due to adsorption of organic molecules	Thiols (R–SH), sulfides (R–S–R), disulfides (R–S–S–R), form SAMs on gold
Glass Silica Metal Oxides	Transparent devices suitable for fluorescence, TIRF, interferometry, etc.	Generally hydrophilic. Mostly terminated with O-bridges and OH groups	Organosilane
Artificial polymers	Integration of optics microfluidics, gratings	From very hydrophobic to very hydrophilic. Native strong adsorption	Wet chemical, flaming or plasma oxidation
Natural polymers	Entrapping matrixes, consumer products	Very hydrophilic, "protein-friendly" nondenaturing	Chloroethylene oxide, trichlorotriazine
Carbon	Microelectrodes. Low-cost disposable e-chemical devices	Very hydrophobic, strong nonspecific adsorption	Wet, electrochemical, or plasma oxidation
Diamond	High-end devices. Long-life, for aggressive chemical environments	Native hydrophobic, slow tendency to hydrophilicity due to surface oxidation	Wet-, electro-, photo-chemical or plasma, amino, ester, alkenes

method and the properties of the bioreceptors to bind. Table 7.1 provides an overview of the most common surface modifications.

Thiols, silanes, and their derivatives are available with many different functional terminations; however, further modifications of the surface properties may be necessary. A wide range of reactions can do these substitutions; some of them are listed in the Table 7.2.

Such modifications of the surface properties need optimization for providing the requested enhancement of performances in the biological environment while retaining the "bulk" properties of the carrier.

7.3 Entrapment

The immobilization of bioreceptor molecules can be realized also by means of entrapment. There are various ways to achieve this goal. In the following, we analyze a few of them.

7.3.1 Entrapment Behind Membranes

The easiest entrapment method is behind a semipermeable membrane, as illustrated below (Fig. 7.4):

Table 7.2 Examples of surface modification reactions [1]

Native surface termination	Reagent	New surface termination
R–NH$_2$	Succinyl oxide (dihydro-2,5-furandione)	R'–C(=O)–OH
R–NH$_2$	Dialdehyde	R'–CH$_2$–CHO
R–(epoxide)–O	Ammonia NH$_3$	R'– NH$_2$
R–(epoxide)–O	Glycine, H$_2$N–CH$_2$–C(=O)–OH	R'–C(=O)–OH
R–SH	Iodoacetic acid, I–CH$_2$–C(=O)–OH	R'–C(=O)–OH
R–OH	Bromoacetic acid, Br–CH$_2$–C(=O)–OH	R'–C(=O)–OH
R–OH	Epichlorohydrin, Cl–CH$_2$–(epoxide)	R'–(epoxide)
R–OH	Trichlorotriazine	R'–(triazine with two Cl)

While small molecules can freely diffuse in both directions through the membrane, large molecules or whole cells used as bioreceptors remain entrapped behind the membrane. The typical molecular weight cut-off is around 5–10 kDa.

Following membrane materials can be used for this function:

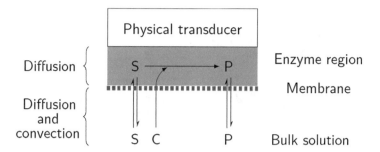

Fig. 7.4 Immobilization behind a semipermeable membrane allows diffusion of small analytes, cofactors and cosubstrates like oxygen, while preventing loss of the large receptor molecules like enzymes. The letters indicate: S = substrate, P = product, C = cofactor and/or cosubstrate

- Cellulose ester (CE)
- Regenerated cellulose (cellophane)
- Polyvinylidene difluoride (PVDF)
- Polyamide (nylon)
- Polyimide (PI, Kapton)
- Teflon (PTFE).

An example showing a practical implementation of this method is the potentiometric biosensor displayed in Fig. 3.16. This method is simple but suitable for single, non-miniaturized sensors only.

7.3.2 Entrapment Within Conducting Polymers

Another approach consists in the immobilization of bioreceptor molecules within electrically conducting polymers. For example, a multi-cofactor enzyme QH-ADH (quinohemoprotein alcohol dehydrogenase) entrapped within polypyrrole (PPy) is able to transfer electrons via the conducting polymer directly to the electrode surface (Fig. 7.5).

In the sensor illustrated above, the following substances are involved:

- Quinohemoprotein alcohol dehydrogenase (QH-ADH): ADH + PQQ + heme c.
- Pyrroloquinoline quinone (PQQ) $C_{14}H_6N_2O_8$, also known as methoxatin, is a natural redox cofactor of alcohol dehydrogenase (ADH).
- The hemoprotein heme c provides direct electrons transfer during the redox reactions.
- Polypyrrole (PPy) is a conducting polymer easily obtained by electropolymerization of pyrrole, directly on the gold electrode biased at the anodic potential of 0.65–0.7 V.

Polypyrrole is prone to degradation, loosing progressively its initial conductivity. In recent years, the use of poly-3,4-ethylendioxythiophene (PEDOT) as electrical conducting polymer has attracted much interest due to its good conductivity and chemical stability in the long term [3, 4]. The polymerization of EDOT monomers also takes place

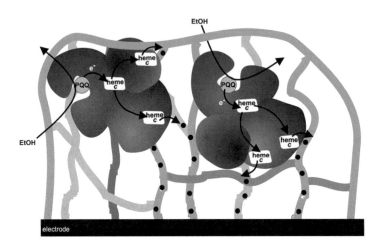

Fig. 7.5 Model illustrating an ethanol sensor based on QH-ADH entrapped in conducting polypyrrole on a gold electrode (Reprinted from [2], © 2000, with permission from The Royal Society of Chemistry)

electrochemically at around 0.7 V. Here, polystyrol sulfonate (PSS) is used instead of chlorine as counter ion. The resulting product is therefore named PEDOT:PSS. Since these polymers are obtained by electrochemical polymerization, they are suitable for electrode arrays, even miniaturized, having several addressable and individually functionalized spots, for multi-analyte detection.

7.3.3 Entrapment in Hydrogel

Another method to immobilize the bioreceptor is its entrapment in a matrix, such as hydrogel. Agarose (algae polysaccharide) is most commonly used for this purpose. When dissolved in hot water, it forms a gel with high porosity when cooling down. Therefore, it is suitable for entrapping large molecules or whole cells, while small molecules can diffuse easily. Entrapped bioreceptors preserve most of their activity. This method is also well suited for designing microarrays (Fig. 7.6).

Alternative hydrogels are polyvinyl alcohol (PVA) or polyacrylamide (PAM). These are synthetic polymers, nearly insoluble in cold water and forming hydrogels with low porosity.

In case of small bioreceptors, which could easily escape from the matrix, they can be anchored to the gel by covalent binding with glutaraldehyde as cross-linker.

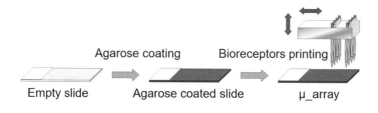

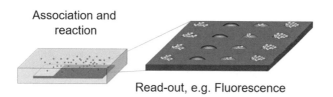

Fig. 7.6 (Top) Creation of a μ-array of bioreceptors on an agarose layer and dispensing of analyte (bottom) followed by detection (Adapted from [5], © 2006, with permission from Elsevier)

Question 3

Which immobilization method through entrapment do you know?
Which characteristics does each method offer?

7.4 Covalent Binding

The third group of methods for immobilizing bioreceptors consists in the formation of covalent bonds. This approach is also called *chemisorption* to distinguish this kind of immobilization from the simpler and weaker *physisorption*, which is based on physical, for example, polar interactions.

Chemisorption is very stable under a wide range of conditions, both chemical (e.g., pH) and physical (e.g., temperature). However due to the strength of covalent bonds, there may appear restrictions in orientation and activity of the immobilized bioreceptors; therefore, adequate strategies are necessary for a good yield.

The following functional groups of bioreceptors and surfaces are most commonly involved:

- Carboxyl:
  R–COOH
- Aldehyde:
  R–CHO
- Thiol:
  R–SH

- Amine:
  R–NH$_2$

Usually, the binding reactions are carried out under mild conditions in water or buffer solutions. Chemistry in organic solvents can also be used, for instance if it is important to prevent oxidation.

7.4.1 Coupling to Carboxylic Acids (Carboxyl Groups)

On a carboxyl-terminated surface, the receptor molecules usually bind with the amino groups present in their structure forming an amide bond, but they can also link to other groups such as phenols or thiols.

By coupling antibodies, the majority of bonds will be with surface amino groups of lysine residues, even if a few surface tyrosine, cysteine, or histidine residues are present since the latter are usually much less abundant.

The amide bond is usually formed in two ways. A first option is by activating the carboxylic group with a carbodiimide reagent like 1-ethyl-3-(3-dimethylpropyl)-carbodiimide (EDC), which then binds to an amine. Alternatively, the carboxylic group is first converted to an "active ester" via N-hydroxysuccinimide (NHS) and an acid chloride, and then the amide is formed again by reaction with an amine (Fig. 7.7).

The two methods described above are most often combined in a single EDC/NHS protocol because in this way the overall yield is significantly enhanced.

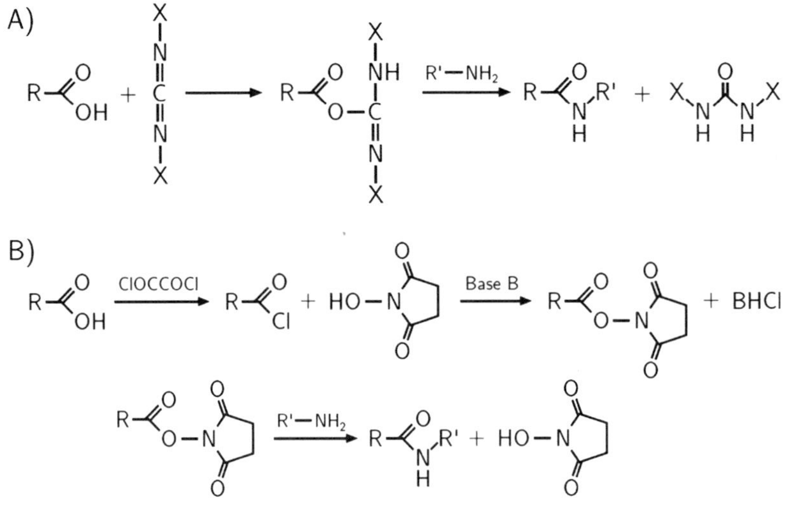

Fig. 7.7 (**a**) Formation of an amide bond by activation of carboxylic acid with EDC, followed by a nucleophilic displacement with the amide. (**b**) Two-step formation of amide bonds via NHS [1]

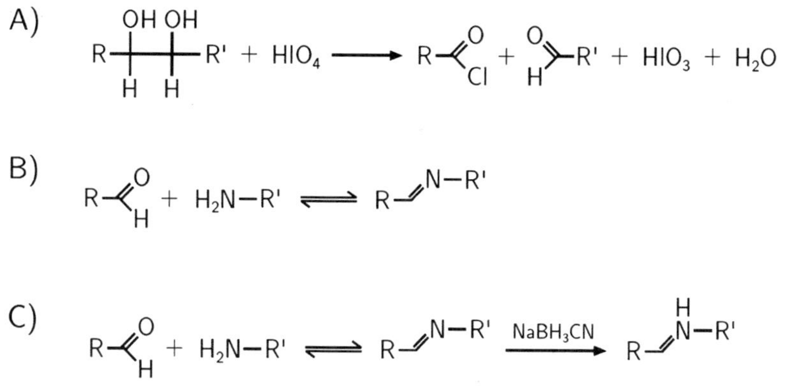

A)

$$R\overset{OH\ OH}{\underset{H\ \ H}{\rule{0pt}{0pt}\vdash\!\vdash}}R' + HIO_4 \longrightarrow R\overset{O}{\underset{Cl}{<}} + \overset{O}{\underset{H}{>}}R' + HIO_3 + H_2O$$

B)

$$R\overset{O}{\underset{H}{<}} + H_2N-R' \rightleftharpoons R\diagup^{N-R'}$$

C)

$$R\overset{O}{\underset{H}{<}} + H_2N-R' \rightleftharpoons R\diagup^{N-R'} \overset{NaBH_3CN}{\longrightarrow} R\diagup^{\overset{H}{N}-R'}$$

Fig. 7.8 (a) Reaction by splitting a vicinal diol (e.g., in a sugar group) with metaperiodic acid to generate aldehyde groups. (b) Formation of an imine followed (c) by an in situ reduction to a secondary amine using sodium cyanoborohydride [1]

7.4.2 Reactions Involving Aldehydes

Aldehyde coupling reactions are of particular interest due to the habitual use of dialdehyde and glutaraldehyde as coupling agent and the ease of preparation of aldehydes.

Due to the position of their carbohydrate chains, the aldehydes produced out of carbohydrate groups of glycoproteins, that is by splitting *vicinal diols*, provide a convenient method for binding antibodies at a defined position and orientation (see Fig. 4.4). Aldehydes react with amines to form imines; however, under typical reaction conditions (aqueous buffer, pH 5–6), this reaction has an equilibrium constant close to 1; therefore, the reaction cannot be completed. Because of this equilibrium, imines are inherently unstable. The problem can be solved by reducing imines to secondary amines (Fig. 7.8).

7.4.3 Reactions Involving Thiols

Free thiol groups are not very stable; in fact, they can be easily oxidized. These oxidations are catalyzed by transition metal ions, so it is generally best to carry out thiol coupling reactions under inert atmospheres and using buffers containing chelators such as ethylenediaminetetraacetic acid (EDTA), which chelates most metals from Ca, Mg, and Fe to lanthanides, Hg, Pb, etc.

Thiol groups play an important role in antibody coupling, due to the possibility of producing Fab' fragments with free thiols. Thiol coupling occurs in general either by disulfide formation or by creating thioethers (Fig. 7.9).

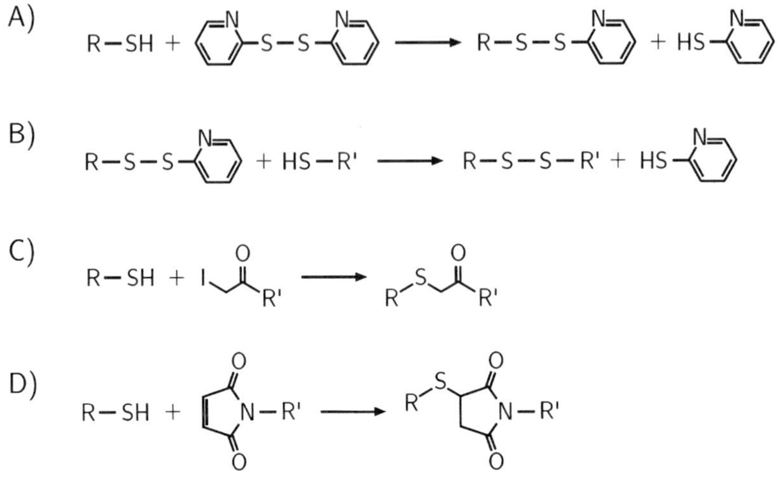

A)

B)

C)

D)

Fig. 7.9 (**a**) Thiol-disulfide exchange between a surface-bound thiol group and a 2,2-′-dithiodipyridine. (**b**) Thiol–disulfide exchange between a disulfide and a ligand with a free thiol groups, leading to ligand immobilization. (**c**) Coupling to thiols via thioether formation using carboxylic acyl iodide and (**d**) maleimide [1]

7.4.4 Reactions Involving Amines

Coupling to amino-terminated surfaces is not widely used, but can give good results for some small ligands. For instance, the carboxylic group of biotin can be directly coupled using the EDC/NHS chemistry seen above, so that the use of expensive biotin derivatives is avoided. The process of covalently binding biotin to a molecule (usually a protein or DNA) is called *biotinylation* (Fig. 7.10).

The big advantage of using biotin as active linker group resides in the possibility of using a crosslinker of the avidin family [6], for forming a "bridge" between biotinylated entities.

Avidin and streptavidin, the most commonly used avidins, are both tetrameric proteins with a high affinity for biotin and four identical coupling sites. This is one of the strongest affinity bonds found in nature with a ratio $K_a/K_d \approx 10^{15}$, that is, de facto irreversible. Furthermore, this is a specific bond; in fact, avidins do not bind easily with other molecules, and if ever with much weaker affinity.

In case the linker to the surface is too short and a denaturation of the bioreceptor becomes a risk, a soft anchoring is possible by using a poly-L-lysine/polyethylene glycol (PLL/PEG) complex (Fig. 7.11).

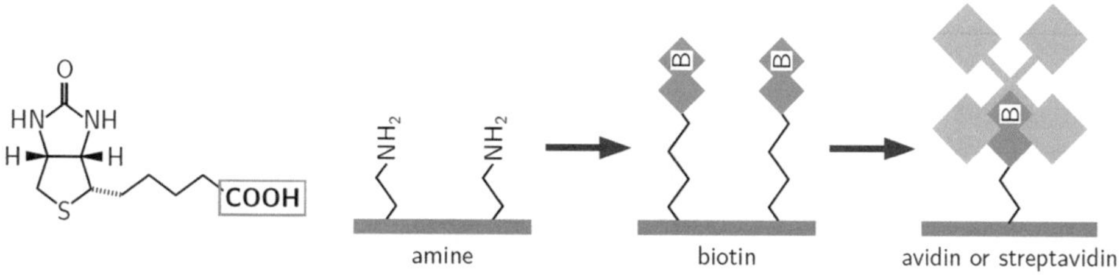

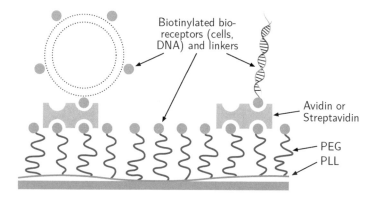

Fig. 7.10 (Left) Structure of biotin. (Right) Binding the carboxylic group of biotin to an amino-terminated surface. The addition of avidin or streptavidin as "crosslinker" results in three identical binding sites available for the attachment of biotinylated bioreceptors

Fig. 7.11 Soft anchoring by PLL/PEG complex prevents denaturation of the bioreceptors

---

Question 4

Which functional groups can be used for the immobilization via covalent binding?

7.5   Self-Assembled Monolayers (SAMs)

The term self-assembled is referring to the spontaneous formation of organized structures or patterns from elementary units. It is widely present in nature (e.g., lipid membrane, cells, organisms). A SAM is an ordered molecular monolayer that forms spontaneously on a suitable surface thanks to its specific affinity with the "head group" of the molecules. The combination of hydrophilic "head group" and a hydrophobic "tail group" is responsible for orientation and organization during the SAM formation. The tail group is at the same time protective against chaining, that is, assuring the growth of a monolayer and preventing a multilayer assembly. In vitro SAM formation is carried out in water-free organic solvents

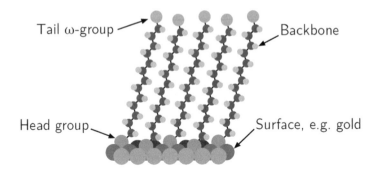

Fig. 7.12 SAM formation on a surface

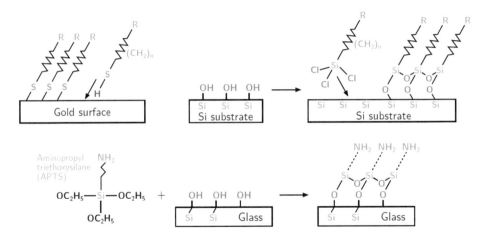

Fig. 7.13 Formation of SAMs: On gold with alkane thiols $HS(CH_2)_nR$ (top left); on silicon with a chlorosilane $SiCl_3(CH_2)_nR$ (top right); on glass and silica (SiO_2) with an organosilane (bottom)

such as anhydrous ethanol (EtOH) or tetrahydrofuran (THF), for preventing oxidation and formation of molecular clusters. In a subsequent step, the *ω-group* (the Greek letter "omega" is used in chemistry for the end group of a chain, i.e., the "tail") can be switched to a different one for a convenient linking to a suitable bioreceptor (Figs. 7.12 and 7.13).

Question 5

What is the cause of the orientation and order of SAMs?

7.5.1 Steric Hindrance

A *steric hindrance* occurs when the size of a molecule hinders the access to reactive sites of nearby molecules, for example, preventing them to react and bind with a linker. It can also

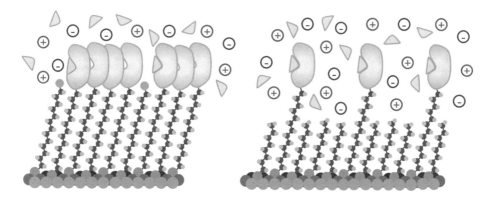

Fig. 7.14 (Left) Effect of steric hindrance: electrolyte and analyte molecules can hardly diffuse into the sensing layer. (Right) Short, nonfunctionalized molecules between the long, functionalized linkers serve as spacers, thus reducing the steric hindrance or shielding

lead to *steric shielding* when the polar groups on nearby molecules interact in such a way that reduces the kinetics of a reaction. This effect can be reduced by adopting either a proper mix of short nonfunctionalized and long-functionalized SA-molecules, or a proper mix of molecules with reactive and nonreactive end groups for the given binding chemistry (Fig. 7.14).

7.6   Long-Term Stability

In the previous observations, we did not consider the long-term stability of the bonds. Figure 7.15 shows the behavior of some carrier materials during several reaction cycles. These behaviors should be carefully considered when selecting the appropriate technology for sensors intended for multiple cycles of utilization and long lifetime.

> **Question 6**
>
> What is the advantage of diamond as carrier material in immobilization?
>   What about gold?

7.7   Immobilization Protocol of a Ligand on a Gold Surface

This section describes the immobilization protocol (Table 7.3) on Biacore Sensor Chip CM5 (Source: *Biacore Sensor Surface Handbook*). It is important to note here that the sensor chip has a two-stage structure for functionalization purposes. Firstly, a nonspecific

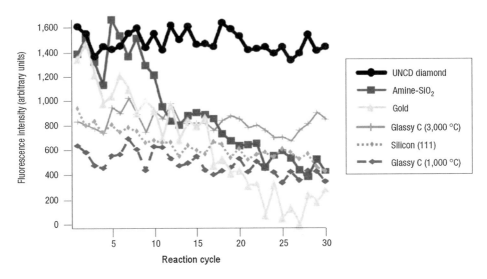

Fig. 7.15 Stability of immobilization bonds on several carrier materials during 30 successive experiments (Adapted from [7], © 2002, with permission from Springer Nature)

Table 7.3 Steps of the amine coupling immobilization protocol

Immobilization step	Chemistry	Parameters
Activation of dextran matrix on the gold surface	EDC/NHS	Flow rate: 5 µL/min Contact time: 10 min
Covalent coupling of the ligand on the dextran matrix	Ligand	Flow rate: 5 µL/min Contact time: 10 min
Deactivation of unbound active groups	Ethanolamine	Flow rate: 5 µL/min Contact time: 7 min

linker layer is anchored on the gold-coated glass slide (e.g., dextran[2] for the CMx sensor, as shown in Fig. 7.16), whereupon a recognition element (usually an antibody) specific to the intended detection is immobilized. This is referred to as ligand. Once the chip is functionalized, it can be used in experiments, in which the ligands detect and bind the analyte molecules in the sample solution (Table 7.3).

For this protocol (Fig. 7.17), the following solutions are required:

- EDC: 0.4 M of 1-ethyl-3-(3-dimethylpropyl) carbodiimide in water
- NHS: 0.1 M N-hydroxysuccinimide in water
- Ethanalomine: 1 M ethanolamine-HCl, pH 8.5
- Ligand: ~200 µL of a solution with a ligand concentration of 50 µg/mL

---

[2] Dextran is a complex polysaccharide chain that consists of many glucose molecules (from 10 to 150 kDa).

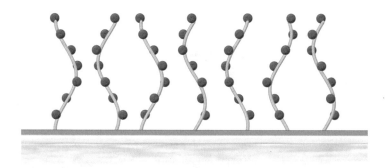

Fig. 7.16 Carboxymethylated ($-CH_2-COOH$) dextran matrix on the surface of Sensor Chip CM5 (© 2020 Cytiva—Reproduced by permission of the owner)

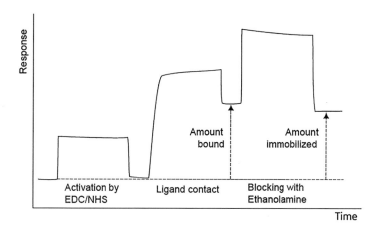

Fig. 7.17 Sensogram of amine coupling procedure, SPR-signal vs. time: Activation: NHS/EDC injection → surface ester, Ligand contact: Reaction with amino group on ligand, Blocking: Deactivation of free esters with ethanolamine. After every step occurs a wash with buffer (© 2020 Cytiva—Reproduced by permission of the owner)

7.7.1   Optimization of the Immobilization

For charged molecules, the binding condition is pH dependent. An efficient pre-concentration requires a buffer with a pH value between the pK_a at the surface and the isoelectric point (pI) of the ligand (Fig. 7.18). A low ionic strength of the buffer is also important for reducing electrostatic shielding. Poor pre-concentration can be partially compensated by increasing the ligand concentration.

pH scouting is the experimental procedure to find the optimal pH value for immobilization. The dextran matrix is negatively charged for pH values above 3.5. The immobilization

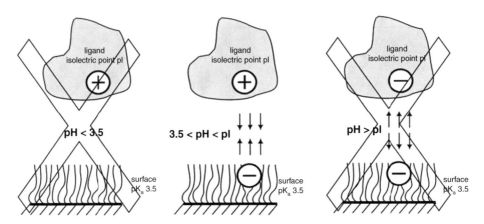

Fig. 7.18 Illustration of the pH influence during the immobilization (© 2020 Cytiva—Reproduced by permission of the owner)

buffer should have a pH value higher than 3.5, but lower than the isoelectric point of the ligand. The isoelectric point (*pI*) of a protein is defined as the pH value at which it possesses no net charge (pH < *pI*: net charge of the protein is positive). For many proteins, a 10 mM sodium acetate buffer (pH = 4.5) works well as a starting point. By performing several immobilization experiments with buffers of different pH values, the optimal yield can be determined. For this purpose, the complete immobilization protocol must be carried out for each pH value.

Figure 7.19 shows the sensograms obtained by pH-scouting with four pH values.

Question 7

In what pH range should be the buffer for the immobilization of a bioreceptor (ligand) on the dextran matrix? Please explain your reasoning.

7.8   Regeneration

An efficient regeneration of the biosensor removes the bound analyte completely. A second injection of the analyte then shows if the ligand is still fully active (Figs. 7.20 and 7.21).

7.8.1   Tuning Optimal Regeneration Conditions

The optimal conditions are specific to the particular ligand–analyte configuration. A fairly narrow range of conditions is generally effective with a wide range of interacting substances.

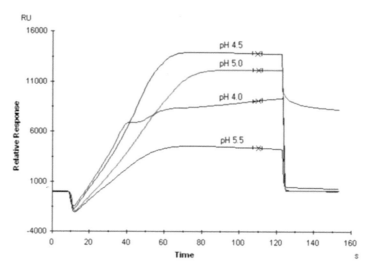

Fig. 7.19 Example of pH scouting. The binding yield increases as the pH is reduced from 5.5 to 4.5. At pH 4.0, the sensogram is irregular and bound material does not dissociate from the surface after the injection, indicating that the protein is aggregating or denaturing. The best results for this protein is at a pH between 4.5 and 5.0 (© 2020 Cytiva—Reproduced by permission of the owner)

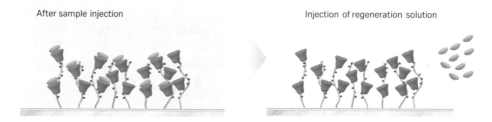

Fig. 7.20 Regeneration of the biosensor by releasing and washing-out the analyte. The ligands remain on chip for the next measurement (© 2020 Cytiva—Reproduced by permission of the owner)

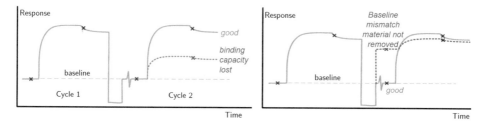

Fig. 7.21 An efficient regeneration removes all bound analyte: Baseline recovery indicates full or only partial analyte removal (left). A second injection of analyte reveals whether the receptor is still fully active (right). Repeated cycles of analyte and regeneration injections are required to evaluate the adopted regeneration conditions (© 2020 Cytiva—Reproduced by permission of the owner)

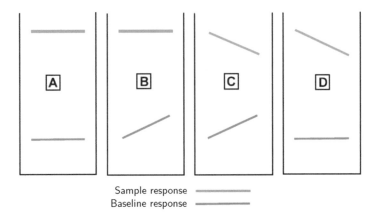

Sample response ‑‑‑‑‑‑‑‑‑‑‑‑
Baseline response ‑‑‑‑‑‑‑‑‑‑‑

Fig. 7.22 Change in sample and baseline responses with repeated trials: (**a**) Ideal case, both baseline and sample response are unaffected by regeneration. (**b**) Baseline shows an upward trend, suggesting accumulation of material on the surface between cycles: However, sample response is unaffected. Regeneration acceptable. (**c**) Baseline upward trend, but with a downward trend in the sample response. Accumulation of material on the surface appears to block analyte binding sites. Regeneration inadequate. (**d**) Constant baseline, but analyte binding capacity is progressively lost, probably because the ligand is damaged by the regeneration conditions. Regeneration inadequate (© 2020 Cytiva—Reproduced by permission of the owner)

Here are suggested alternative starting points for protein ligands:

1. Low pH (10 mM glycine-HCl, from pH 3 to pH 1.5)
2. Ethylene glycol (50%, 75%, and 100%)
3. High pH (1–100 mM NaOH)
4. High ionic strength (up to 5 M NaCl or 4 M MgCl2)
5. Low concentrations of the surfactant sodium lauryl sulfate (SLS) (up to 0.5%)
   - *Ideal regeneration*: The analyte signal is consistent after repeated injections and within 10% of the level of the first injection (Fig. 7.22A).
   - *Too-mild conditions*: The analyte signal decreases and the baseline response increases (Fig. 7.22C).
   - *Too-harsh conditions*: The analyte signal decreases and the baseline response remains constant or decreases (Fig. 7.22D).

Question 8
What happens if conditions are too mild or harsh when regenerating a biosensor?

Answer 1

There are a few strategies for preventing the steric deformation of bioreceptors:

- Use fitting molecules, which hold the binding site in the proper shape during the immobilization.
- Use coupling methods, which bind under the native, energetically most convenient form, since any deformation is linked to an energy shift due to tension or compression.
- Use spacer molecules, tethers, which provide an appropriate clearance from the carrier surface, thus reducing its steric influence.

Answer 2

Adsorption, the simplest immobilization method, is characterized by the physical forces (hydrogen bond, ionic bond, and Van der Waals forces) between the carrier surface and the receptors. It is therefore unspecific and other unwanted molecules can bind as well, thus generating artifacts. Receptor orientation is random and denaturation can be observed especially on hydrophobic surfaces due to strong interaction with the hydrophobic domains of the proteins. Furthermore, loss of receptors can occur when changing temperature or pH. Best results are achieved under mild aqueous conditions mimicking the natural environments. This method is best suited for cheap, single-use devices.

Answer 3

1. Behind membranes: This is an old, simple, macroscopic approach, which makes use of polymeric porous membranes. Here ions and small molecules like analytes and reaction products can diffuse in and out, while large molecules like enzymes remain entrapped in the sensing region. This method is simple but suitable for single, non-miniaturized sensors only.
2. In a conducting polymer: Polypyrrole and PEDOT:PSS are examples of polymers having a 3D porous structure which allow diffusion of small molecules and ions, while holding large biomolecules in place without denaturation. Suitable cross-linkers can provide direct electron transfer to the electrode, thanks to the conductivity of such polymers. Since these polymers are obtained by electrochemical polymerization, they are suitable for arrays with several addressable spots having individual functionalization.
3. In hydrogel: Similar to the above, regarding the diffusion and entrapping properties. Gel is obtained after cooling a hot water–based solution of agarose or polyvinyl alcohol or polyacrylamide. It is typically created on glass or other non-conducting carriers and therefore fluorescence detection is commonly adopted. This method allows arraying as well, multiple functionalization can be provided manually by pipetting or with a multi-needle printing head.

Answer 4
Most common are carboxylic groups, amino groups, aldehydes, and thiols.

Answer 5
Suitable molecules have a hydrophilic head and a hydrophobic tail. The head binds to a suitable surface, while the tail points toward the bulk solution. In this way, all molecules are oriented the same way and form a monolayer, since the tail bears a protective group, which prevents chaining.

Answer 6
Diamond offers the most stable immobilization based on the covalent carbon–carbon bond. Therefore, biomolecules immobilized on diamond can be used for many more measurements than any other carrier material, making this technology particularly interesting for professional uses. Gold is very attractive due to the affinity to thiols; however, this bond is less stable and a progressive loss of receptors is observed.

Answer 7
It depends on the isoelectric point of the receptor. The pH should be above 3.5 (*pI* of dextran) and below the *pI* of the receptor. In this way, the dextran matrix is negatively charged and the receptors are positively charge, thus being attracted toward each other.

Answer 8
Too mild conditions do not remove all analyte molecules; therefore, the next measurement will have a raised baseline and a limited dynamic range. Too harsh conditions lead to loss of receptors, resulting in a constant or lower baseline and a lower response for the same analyte concentration.

Take-Home Messages
- In immobilization, we usually aim for stable binding, retention of function, high signal-to-noise ratio, and low background signal.
- Affinity is simple, cheap, but also unspecific and not very stable; moreover, it can lead to denaturation and unpredictable orientation, with reduction of function.

(continued)

- Entrapment methods provide a good and stable immobilization of bioreceptors. The fabrication effort is moderate. Some methods are suitable for arraying.
- Covalent attachment methods are the most stable ones. They require generally complex chemical procedures. Suitable for reusable protocols based on expensive receptors.
- Biotin–Avidin binding is convenient for immobilization of large molecules and even cells. Even if it is on affinity basis, it is as stable as a covalent bond.
- SAMs are helpful for anchoring receptors at a proper distance from the carrier surface, thus avoiding related steric effects.
- Immobilization and regeneration conditions, like pH and concentration of reagents and buffers, must be properly tuned for reliable and reproducible results.
- The table summarizes the pros and cons of the discussed immobilization methods (Table 7.4).

Table 7.4 Summary of the different immobilization methods

Method	Advantages	Disadvantages
Adsorption	High capacity, simple, inexpensive, good for single use	Protein denaturation, relatively unstable, random orientation
Entrapment in hydrogel: polyacrylamide, agarose	No modification, high capacity, low background	Random orientation, denaturation by free radicals, mass production potential
Entrapment behind membrane: cellulose, cellophane, nylon, Teflon	Simple universal approach for cells and macromolecules	Difficult to mass produce, slow diffusion, large response time
Covalent: carboxyl, aldehyde, amine, epoxy, thiol	Stable attachment, high sensitivity, uniform orientation, fast response	Protein modified, complex, and expensive, steric hindrance
Affinity: biotin/streptavidin	Strong attachment, low background, uniform orientation	Protein modified (biotinylated), expensive, steric hindrance

References

1. Lowe CR, Gizeli E, editors. Biomolecular sensors. New York: Taylor & Francis; 2002.
2. Schuhmann W, Zimmermann H, Habermüller K, Laurinavicius V. Electron-transfer pathways between redox enzymes and electrode surfaces: reagentless biosensors based on thiol-monolayer-bound and polypyrrole-entrapped enzymes. Faraday Discuss. 2000;116:245–55.

3. Jonas F, Schrader L. Conductive modifications of polymers with polypyrroles and polythiophenes. Synth Met. 1991;41(3):831–6.
4. Heywang G, Jonas F. Poly(alkylenedioxythiophene)s—new, very stable conducting polymers. Adv Mater. 1992;4(2):116–8.
5. Mant A, Tourniaire G, Diaz-Mochon JJ, Elliott TJ, Williams AP, Bradley M. Polymer microarrays: identification of substrates for phagocytosis assays. Biomaterials. 2006;27(30):5299–306.
6. Taskinen B, Airenne TT, Jänis J, Rahikainen R, Johnson MS, Kulomaa MS, et al. A novel chimeric Avidin with increased thermal stability using DNA shuffling. PLoS One. 2014;9(3):e92058.
7. Yang W, Auciello O, Butler JE, Cai W, Carlisle JA, Gerbi JE, et al. DNA-modified nanocrystalline diamond thin-films as stable, biologically active substrates. Nat Mater. 2002;1(4):253–7.

Biochips

8

Contents

Keywords

DNA · RNA · Genes · Gene expressions · Protein synthesis · Restriction enzymes · Amplification · Gel electrophoresis · Arraying · Clustering

What You Will Learn in This Chapter
In this chapter, we will deal with biochips. You will first get an overview of DNA, RNA, genes, and gene expression as well as methods of DNA amplification. In the next parts, you will get to know DNA chips, different methods of their functionalization, and the description of their use as diagnostic tools.

You will be able to explain important methods of DNA amplification, to suggest methods of chip functionalization for the respective applications, and to name appropriate evaluation methods.

Biochips, also called *microarrays*, build the basis for miniaturized biochemical assays. Although biochips are often associated with DNA chips, different chemical/biochemical receptors can be arrayed on one chip. This is for instance the case of protein microarrays, which will be described in Chap. 10.

The most significant advantages over conventional methods are as follows:

- A variety of analytes can be detected simultaneously in the same sample.
- The required sample quantities are minimal.
- Low consumption of scarce reagents.
- High miniaturization.
- High sample throughput.

The drug screening represents an example application of this approach. The aim is, in this case, to identify one compound or a combination of compounds out of millions of compounds, as a suitable medication for a specific purpose. The steps can be summarized as follows (Fig. 8.1):

1. A whole chemical library (millions of different molecules) is tested against molecular targets.
2. Selected molecules from (1), hundreds to thousands, are tested with cell models.
3. Selected molecules from (2), a few tens, are tested in animal models.
4. Selected molecules from (3), a few units, are tested with human volunteers in clinical trials.
5. The most successful medication undergoes a market approval procedure (European Medicines Agency [EMA], United States Food and Drug Administration [FDA], etc.).
6. After approval the new drug is released to the market [1–5].

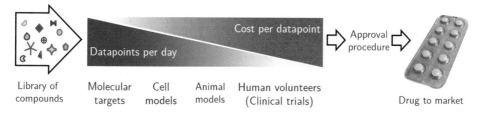

Library of compounds | Molecular targets | Cell models | Animal models | Human volunteers (Clinical trials) | Drug to market

Fig. 8.1 The step-by-step process of drug screening

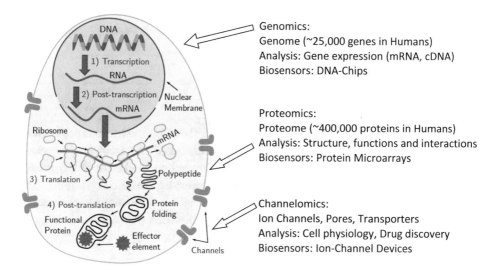

Genomics:
Genome (~25,000 genes in Humans)
Analysis: Gene expression (mRNA, cDNA)
Biosensors: DNA-Chips

Proteomics:
Proteome (~400,000 proteins in Humans)
Analysis: Structure, functions and interactions
Biosensors: Protein Microarrays

Channelomics:
Ion Channels, Pores, Transporters
Analysis: Cell physiology, Drug discovery
Biosensors: Ion-Channel Devices

Fig. 8.2 Scheme of a eukaryotic cell showing the protein synthesis pathway in four steps: (1) Gene transcription from DNA and synthesis of RNA; (2) post-transcriptional splicing and release of mRNA; (3) translation, i.e., synthesis of a polypeptide; (4) post-translational modification and release of a functional protein. Membrane channels are a specific class of proteins. These biomolecular processes are investigated in the respective "omics", i.e., disciplines linked to specific molecular families

In the above procedure, which takes in average between 10 and 15 years, the data points per day are extremely large at the beginning and progressively decrease with increasing model complexity.

In order to understand how such procedures work, we need to discuss the basic concepts of molecular biology. We start looking at the pathway of protein synthesis in a cell, as illustrated in the following cartoon of a eukaryotic cell (i.e., a cell with nucleus). At the different stages, different families of molecules are related to the respective disciplines, with their specific goals and tools (Fig. 8.2).

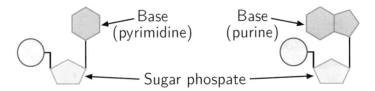

Fig. 8.3 DNA consists of nucleotides. The four nitrogenous bases A, C, G, and T are classified in the two families of purines and pyrimidines

---

Question 1

Name the advantages of biochips compared to conventional methods.

8.1 DNA

The genetic code of all living organisms is stored inside the cells, in eukaryotic cells – more precisely in their nucleus – in form of deoxyribonucleic acid (DNA). DNA is a long chain molecule composed of four different building blocks, called *nucleotides* or simply *bases*). Each nucleotide consists of one unit of deoxyribose (pentose), one phosphate group, and one nitrogenous base (out of the possible four ones), which is the distinctive character of the specific nucleotide and the carrier of the genetic information (Fig. 8.3).

There are four nitrogenous bases:

- Purines: Adenine (A), Guanine (G)
- Pyrimidines: Cytosine (C), Thymine (T)

By means of phosphodiester bonds, the nucleotides form linear strands, called *polynucleotides* (short chains, $n < 100$, are called *oligonucleotides*). In its normal state, DNA is organized in the form of a double helix of complementary nucleic acids, that is complementary strands (Fig 8.4). The DNA stores the instructions for building the whole variety of proteins of a living organism (proteome). The human genome consists of about 3×10^9 nucleotides, which would be about 1 m long if arranged in a linear chain.

Question 2

Why is there no third H-bond between A and T? Why can T and G not pair?

8.1.1 Genes

A **gene** is a region of DNA that controls a hereditary feature. It consists of exons (code sequences for specific proteins) and introns (regulatory sequences without proteinogenic

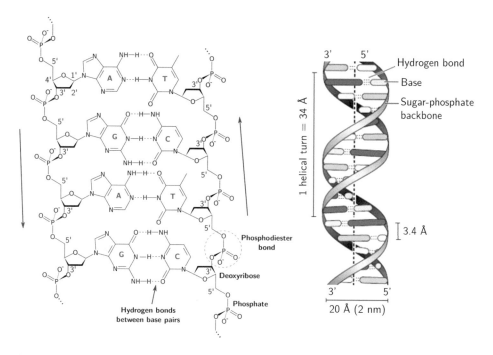

Fig. 8.4 (Left) DNA structure, consisting in two antiparallel complementary strands. The carbon atoms of a deoxyribose ring are numbered from 1′ to 5′. The covalent phosphodiester bond is located between 3′ and 5′ of two adjacent sugars. Due to constraints in the width of 20 Å, the bases pair one purine and one pyrimidine with H-bonds: A and T pair with two H-bonds, while C and G form three H-bonds. The arrows show the growth direction during the DNA synthesis. (Right) Double helix structure of DNA (Adapted from [6], © 2006, with permission from Elsevier)

information). The length of a gene varies between thousands and hundreds of thousands bases. Human DNA consists of approximately 2.5×10^4 genes.

When the genetic information of a gene is utilized for synthesizing a protein, we speak of "gene expression" (Figs. 8.5, 8.6 and 8.7).

8.1.2 Genetic Code

Each amino acid is coded by a triplet or "codon" (3-base sequence).

There are several codons for the same amino acid (Table 8.1). Three codons TAA, TAG, and TGA code the "STOP," that is, the protein end. Although ATG codes methionine, it can also indicate the "START" of the coded protein.

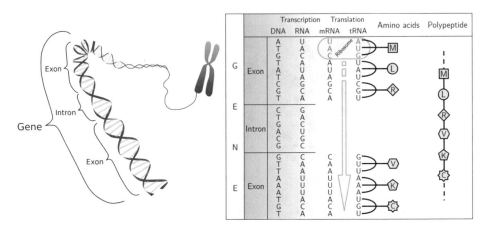

Fig. 8.5 (Left) In the cell nucleus the DNA is tightly coiled in chromosomes; genes consist of exon and intron regions. (Source [7], © Smedlib, CC-BY-SA 4.0 [8]). (Right) Exons contain the code for building polypeptide chains, while introns have regulatory functions. The mRNA consists of complementary triplets called **codons**. Every single tRNA molecule has a domain called *anticodon*, which is complementary to the related codon

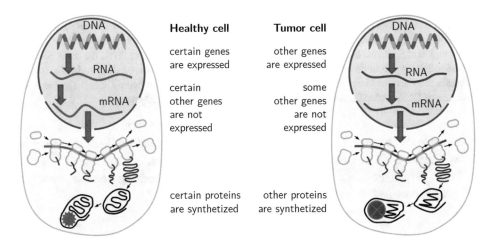

Fig. 8.6 Identification of genes expressed under normal and abnormal conditions

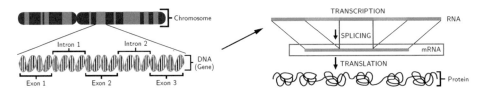

Fig. 8.7 Synthesis of mRNA when a chromosomal gene is expressed for protein synthesis. The mRNA is the target molecule for the analysis of gene expression

Table 8.1 Genetic coding (triplets), at DNA level, of the 20 proteinogenic amino acids

First base	Second base								Third base
	T		C		A		G		
T	TTT	Phe (F)	TCT	Ser (S)	TAT	Tyr (Y)	TGT	Cys (C)	T
	TTC		TCC		TAC		TGC		C
	TTA	Leu (L)	TCA		TAA	STOP	TGA	STOP	A
	TTG		TCG		TAG		TGG	Trp (W)	G
C	CTT		CCT	Pro (P)	CAT	His (H)	CGT	Arg (R)	T
	CTC		CCC		CAC		CGC		C
	CTA		CCA		CAA	Gln (Q)	CGA		A
	CTG		CCG		CAG		CGG		G
A	ATT	Ile (I)	ACT	Thr (T)	AAT	Asn (N)	AGT	Ser (S)	T
	ATC		ACC		AAC		AGC		C
	ATA		ACA		AAA	Lys (K)	AGA	Arg (R)	A
	ATG	Met (M)	ACG		AAG		AGG		G
G	GTT	Val (V)	GCT	Ala (A)	GAT	Asp (D)	GGT	Gly (G)	T
	GTC		GCC		GAC		GGC		C
	GTA		GCA		GAA	Glu (E)	GGA		A
	GTG		GCG		GAG		GGG		G

8.2   Applications of Functional Genomics

One of the most important applications of functional genomics is the identification of complex genetic diseases. The gene expression of a cell in a pathologic condition is different from that of a healthy cell (Fig. 8.6). The analysis consists of comparing the genes that have been expressed and those that have not been expressed. This is helpful in detecting and classifying a disease (e.g., cancer), with the goal of developing a tailored therapy against such disease.

Additional applications are follows:

- Drug development
  - Genotype specific
  - Addressing the cause of a disease better than just its symptoms
- Assessment of mutations/polymorphisms
- Genetic tests (e.g., forensic identification)

8.2.1 Gene Expression

The RNA transcript of a DNA segment belonging to a gene is called *messenger* RNA (mRNA). Figure 8.7 shows the steps for synthesizing the mRNA. The *precursor mRNA* (or simply RNA) is generated as a direct copy of the gene (DNA) by *transcription*. Afterwards, the intron sequences are removed by *splicing*. The finished mRNA can finally leave the nucleus and be used for protein synthesis.

For examining the gene expression, it is necessary to extract the mRNA from the cells. The following steps are carried out in this procedure:

- Breaking-up the cell (cell lysis): mechanically, enzymatic, laser-induced cell lysis
- Precipitation
- Purification

Question 3
What is mRNA? How is it extracted in the case of eukaryotes?

8.3 RNA

Let us now give a closer look at RNA. Its full name is *ribonucleic acid*; therefore, it is a nucleic acid and consists of a single strand of many nucleotides. The primary functions of RNA are related to the transfer of genetic information and transport of amino acids for the synthesis of proteins. RNA is therefore available in different forms, like mRNA that acts as an information transmitter and tRNA that transports coded amino acids to the ribosomes for building the peptide chains.

RNA has the same primary structure as DNA, with the following exceptions (Fig. 8.8):

- It always exists as a single strand.
- An OH group is present at position $2'$ of the sugar, that is, ribose instead of deoxyribose.
- The base uracil (U) replaces thymine (T), the difference is one methyl-group missing.
- It shows secondary and tertiary structures due to hybridization of complementary bases (folding and looping onto itself).

8.3.1 Ribosomes and Protein Synthesis

The protein synthesis begins in the ribosomes and starts when an mRNA molecule arrives. Several *transfer RNA* (tRNA) molecules are involved in this process. Every tRNA possesses a domain with the specific *anticodon* matching a specific triplet of the mRNA

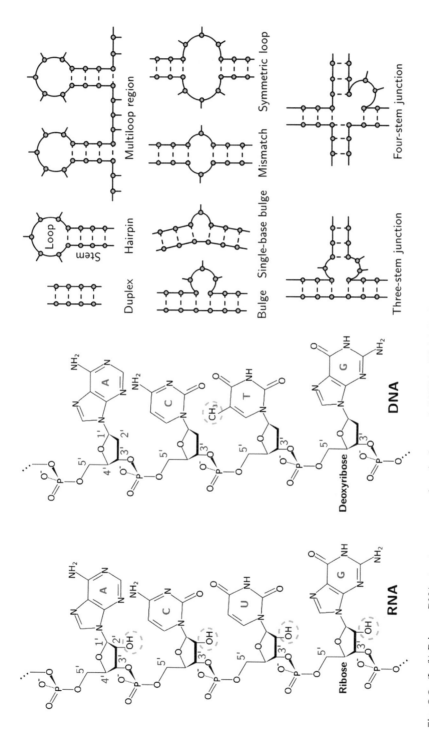

Fig. 8.8 (Left) Primary RNA structure compared to single-stranded DNA. (Right) Basic shapes of RNA secondary structure

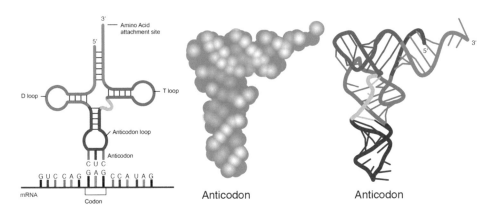

Fig. 8.9 (Left) "Flattened" representation tRNA. (Right) Corresponding 3D illustrations of folded tRNA (Courtesy: National Human Genome Research Institute, www.genome.gov)

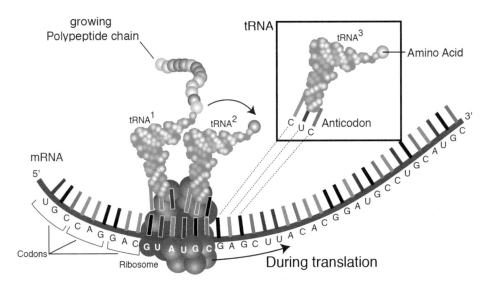

Fig. 8.10 Synthesis of the polypeptide chain (Courtesy: National Human Genome Research Institute, www.genome.gov)

and carries at the 3′ end the corresponding amino acid (Fig. 8.9). The start signal is the triplet ATG, which codes methionine. When a STOP triplet shows up, no more tRNA is bound and the polypeptide chain is released (Fig. 8.10).

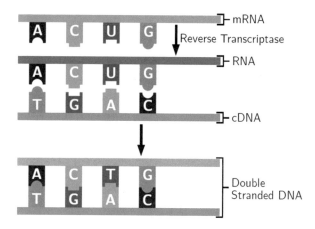

Fig. 8.11 Production of double stranded DNA by reverse transcriptase from mRNA

8.4   cDNA

The mRNA molecules, obtained after the purification step, are less stable compared to DNA (temperature, oxidation, etc.). Thus, the *complementary* DNA *(cDNA)*, which is an exact copy of the gene transcribed into the mRNA, is of practical importance. cDNA is obtained by *reverse transcriptase*. The synthesis of the second strand of DNA delivers a double-stranded DNA which is a clone of the expressed gene (Fig. 8.11).

8.5   Amplification

In the next step, the gene can be amplified by *cloning vector* or *PCR* and then used for bioanalysis.

Restriction enzymes are involved in the amplification with cloning vector. These enzymes are isolated from bacteria and are able to cleave ("digest") DNA at specific sites (sequences of 4 up to 12 or more bases). For this, the restriction enzymes detect specific sequences of the DNA. For instance, the enzyme EcoRI[1] (eco R one) recognizes the sequence GAATTC. So far, more than 50,000 restriction enzymes with hundreds of different specificities have been isolated [9, 10].

Restriction enzymes can be classified as follows:

* *Endonucleases* are enzymes, which cleave within the molecule.
* *Exonucleases* cleave at one end of the molecule.

---

[1] Restriction enzymes are named after the bacteria from which they are isolated, for example: EcoRI = *genus* **E**scherichia, *species* **co**li, *strain* RY13, *order of identification* I, SmaI = *genus* Serratia, *species* marcescens, *order of identification* I.

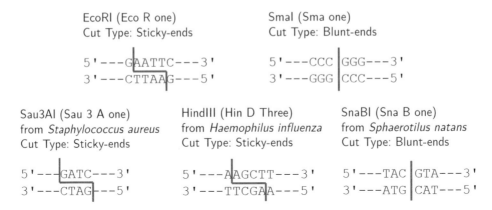

EcoRI (Eco R one)   SmaI (Sma one)
Cut Type: Sticky-ends   Cut Type: Blunt-ends

```
5'---G AATTC---3'              5'---CCC GGG---3'
3'---CTTAA G---5'              3'---GGG CCC---5'
```

Sau3AI (Sau 3 A one)   HindIII (Hin D Three)   SnaBI (Sna B one)
from *Staphylococcus aureus*   from *Haemophilus influenza*   from *Sphaerotilus natans*
Cut Type: Sticky-ends   Cut Type: Sticky-ends   Cut Type: Blunt-ends

```
5'--- GATC---3'        5'---A AGCTT---3'        5'---TAC GTA---3'
3'---CTAG ---5'        3'---TTCGA A---5'        3'---ATG CAT---5'
```

Fig. 8.12 Examples of restriction enzymes

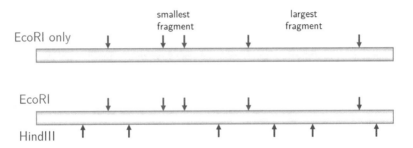

Fig. 8.13 Digestion of a DNA fragment with endonucleases: (Top) A linear DNA molecule with five "GAATTC" regions will be cut into six fragments of various length. (Bottom) The combination of two enzymes (e.g., EcoRI + HindIII) allows isolating different regions of one molecule, thus a more specific target sequence can be found. A given DNA or RNA molecule will generate a characteristic series of patterns when digested with a set of different enzymes

- *DNAses* are specific for DNA.
- *RNAses* are specific for RNA.
- *Type of cut* are "sticky ends," that is, with open H-bonds, or "blunt ends" (Fig. 8.12).

Different enzymes recognize their specific target sites with different frequencies:

- EcoRI recognizes an hexameric (six bases) sequence → probability 4^{-6}.
- Sau3A1 recognizes a tetrameric (for bases) sequence → probability 4^{-4}.

Thus, Sau3A1 cuts the same DNA molecule more frequently. Two or more restriction enzymes can be used in a certain protocol for recognising DNA regions with specific sequences (Fig. 8.13).

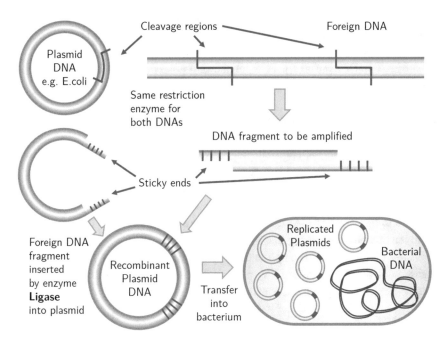

Fig. 8.14 Cloning vector: Generation of recombinant DNA and amplification by autonomous intracellular plasmid replication

8.5.1   Amplification by Cloning Vector

After restriction, a *recombinant* DNA can be generated. For this purpose, the foreign polynucleotide to be amplified is inserted into the ring-shaped extrachromosomal DNA (plasmid) of a bacterium (e.g., *Escherichia coli*), with the help of the enzyme **ligase**. The plasmid will then replicate autonomously inside the cell producing up to several hundred copies (Fig. 8.14). Additionally, the whole bacterium replicates several times. Both processes yield the requested amplification of the plasmid. The foreign DNA segment (of interest) can then be isolated by means of restriction enzymes after cells lysis and purification.

8.5.2   Amplification by PCR

The polymerase chain reaction (PCR) is a purely enzymatic amplification method [11]. It is suitable for amplifying relatively short, precisely defined fragments of DNA.

In order to run a PCR process, you need:

- The original DNA that contains the fragment(s) to be amplified (template)
- Two primers to determine the starting point of the DNA synthesis on both single strands

- A mix of all four bases, in form of dNTPs (deoxyribonucleotide triphosphate)
- The thermostable enzyme *Taq* DNA polymerase from the bacterium *Thermus aquaticus*
- Magnesium Ions (Mg^{2+}) as cofactor for the enzyme polymerase
- Optimized Tris-HCl buffer for PCR, containing potassium (K^+) or better ammonium (NH^{4+}) ions, usually at a pH between 8.0 and 9.5

The PCR consists of the following steps (Fig. 8.15):

1. *Denaturation* (melting) at approximately 94–96 °C: The hydrogen bonds become so weak that the double stranded structure splits in two single strands.
2. *Primer annealing* at approximately 54–57 °C: Hydrogen bonds between primer molecules and the template are constantly formed and broken (the primer is the starting structure to be amplified). At the positions where primers exactly fit the template, the hydrogen bonds are so strong that the primers stays attached.
3. *Extension* (polymerization, amplification) at approximately 72 °C: The enzyme polymerase adds dNTP's from $5'$ to $3'$, reading the template from $3'$ to $5'$. The bases are added complementary to the template.

The above steps are repeated for 20–30 cycles, in some cases up to 40 times, in an automated "Thermo Cycler".

Every cycle doubles the number of the DNA strands in the reaction vessel. This results in an exponential amplification of the DNA present between the primers. The total amplification obtained at the end of the process is given by 2^n; where n is the number of PCR cycles performed. After 20 cycles, this yields approximately an amplification of 1×10^6 and after 40 cycles the amplification would be approximately 1×10^{12}.

Fine-tuning of all parameters and precise control of the process are particularly important, since exponential amplification can also lead to exponential increase of errors.

Question 4

Explain the PCR! Which steps are necessary?

8.5.3 Gel Electrophoresis

Gel electrophoresis is a widely used technique for analyzing *nucleic acids* and *proteins*. Agarose gel is routinely used for DNA analysis while polyacrylamide is more commonly used for proteins and short DNA strands.

Since DNA is negatively charged, it will migrate toward the anode in an electrical field (Fig. 8.16). The gel slows by "friction" the migration velocity of the DNA molecules. This migration velocity of DNA is inversely proportional to the $\log_{10}$ of molecular weight, so that DNA separates by size. At the end of the process, DNA is detected by radioactive

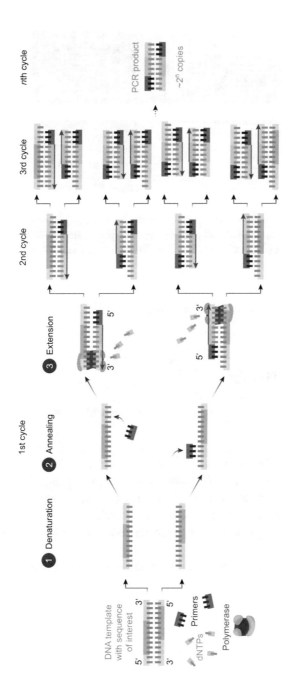

Fig. 8.15 Workflow of polymerase chain reaction. The three-step cycle is usually repeated 20–30 times (Source [12],

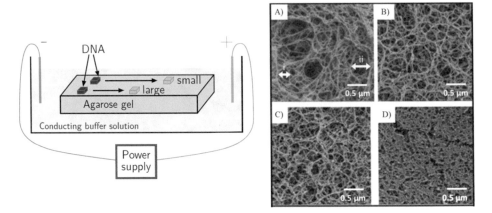

Fig. 8.16 (Left) Gel electrophoresis. (Right) SEM images, showing hydrogel porosity obtained with
(**a**) 0.5%, (**b**) 2%, (**c**) 4%, and (**d**) 8% agarose. Agarose pores allow DNA migration with size-
dependent velocity (Adapted from [13], © 2012 by the authors, license MDPI, CC BY 3.0 [8])

labeling or fluorescence staining (in most cases, ethidium bromide is used as a dye in gel, in
buffer, in sample, or in immersion solution after the run).

8.5.4   PCR Parameter Optimization

Several parameters need optimization for an accurate amplification. Among others, we can
mention the following:

- *Annealing temperature*: The primers are usually designed beforehand on a model basis
  and have a pre-calculated annealing temperature of, for example, 56 °C. However, the
  process should be verified experimentally, for example, in steps of a few °C. If the
  temperature is too low, the PCR yields a mix of specific and unspecific products. If the
  temperature is too high, the primer will not hybridize and no amplification takes
  place (Fig. 8.17 left).
- *Denaturation temperature*: Lower value than the recommended one (e.g., 92 °C) results
  in poor amplification.
- Mg^{2+} *concentration*: The PCR fidelity also depends on the concentration of this ion. It
  can be optimized by variations in steps of 0.5 mMol. Sometimes a compromise between
  yield and specificity is required.
- DNA *Template concentration*: It is source dependent, for instance mammalian genomic
  DNA may vary between 100 and 250 ng for a 50 µl PCR-tube, while for bacterial
  plasmids 20 ng can be sufficient for the same reaction volume.

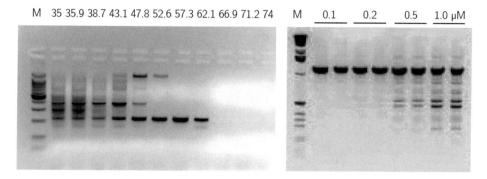

Fig. 8.17 Control by gel electrophoresis after PCR: DNA ladders (references) are shown in the column labeled "M". (Left) Annealing temperature optimization, below 57 °C results in a mixture of specific and non-specific products, between 57.3 and 62.1 °C yields a single specific PCR product, while above 62.1 °C amplification ends because primers are unable to hybridize to the template DNA. (Adapted from [14], © 2008, with permission from Springer Nature). (Right) An excess of primer concentration results in nonspecific amplification products (Adapted from [15], © 2008, with permission from Springer Nature)

- *Primer concentration*: Optimal values range usually between 0.1 and 0.5 μM. Lower values lead to very low efficiency, while higher concentrations result in mispriming and therefore in nonspecific amplification products (Fig. 8.17 right).
- *Concentration of dNTPs*: Optimal values are around 200 μM for a 50 μl tube. Higher values hinder the PCR process.
- Under overall optimal conditions, the *TAQ* DNA polymerase has a base error rate below 10 ppm.

8.6   Southern Blot

Blotting is the technique of transferring DNA to a solid support after the gel electrophoresis.

The Southern Blot (named after its inventor Edwin M. Southern) takes advantage of the fact that DNA fragments will stick to a nylon or nitrocellulose filter membrane. Such membrane is laid over the agarose gel and absorbent, material, that is, blotting paper, is placed on top. With time, the DNA fragments travel from the gel to the membrane by capillary action as surrounding liquid is drawn up to the absorbent material. After the transfer of DNA fragments has occurred, the membrane is washed and then the DNA fragments are permanently fixed to the membrane by heating or UV light exposure. The membrane is now an image of the agarose gel. Finally, the DNA fragments are hybridized with sequences, labelled with radioisotopes, complementary to the ones whose presence needs to be assessed and then labels are detected by exposure to X-ray film, delivering an *autoradiogram* (Fig. 8.18).

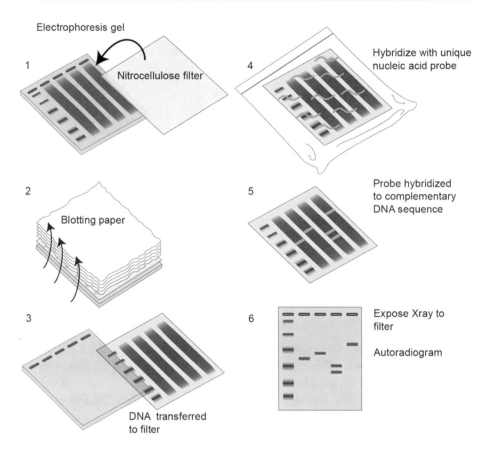

Fig. 8.18 Steps of the Southern Blot procedure (Courtesy: National Human Genome Research Institute, www.genome.gov)

8.6.1 Example: DNA Fingerprinting

DNA fingerprinting is a laboratory technique used to establish a link between two biological samples, for example, in a criminal investigation. A DNA sample (evidence) taken from a crime scene is compared with a DNA sample from a suspect. If the two DNA profiles match, then the evidence came from that suspect. Conversely, if the two DNA profiles do not match, then the evidence cannot have come from the suspect (Fig. 8.19). DNA fingerprinting is also used for paternity testing.

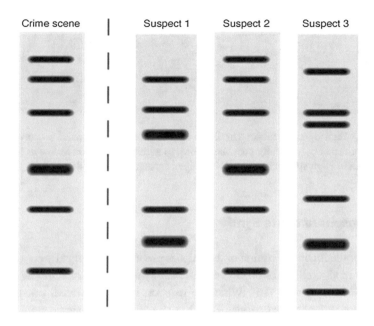

Fig. 8.19 DNA fingerprinting (Courtesy: National Human Genome Research Institute, www. genome.gov)

8.7 DNA Chips

DNA chips are usually available in form of arrays, which are functionalized at different spots with different molecules (see Fig. 2.3). Hybridization, that is, formation of double strands, takes place at spots with matching, that is, when analyte strands result complementary to the functionalization present on that spot. The readout is performed most commonly by laser-induced fluorescence (LIF).

DNA chips have application-dependent requirements. They are expected to respond proportionally to the degree of hybridization and be highly specific. In other words, the dynamic range should be wide and linear with limit of detection as low as just one single hybridization, and reject all mismatches even of a single base in any one strand (Fig. 8.20).

Applications are for example:

- gene expression: drug screening
- medical diagnosis: SNP[2] detection, cancer diagnosis

---

[2] SNP: single nucleotide polymorphism, that is, single nucleotide variation in a given gene.

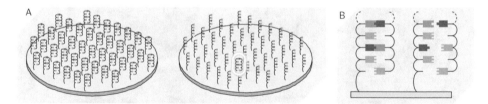

Fig. 8.20 (a) High dynamic range: The DNA chip is expected to respond proportionally to the degree of hybridization, from fully populated, down to a limit of detection as low as just one single hybridization. (b) High specificity: The DNA chip is expected to reject strands with just one single base mismatch

8.7.1 Phosphoramidite Synthesis

The DNA for the functionalization of the chips is produced synthetically on a substrate by means of an in vitro chemical process named *beta-cyanoethyl-phosphoramidite* or simply *phosphoramidite* synthesis [16]. When the reaction is run in a tube, microspheres are commonly used as substrate (carrier). They have a high surface/volume ratio, which leads to a high product density and requires small amounts of expensive reagents.

The steps of this procedure, illustrated in Fig. 8.21, are the following:

1. Start: Microsphere carry linkers with the first base, protected with DMTr at 5'.
2. Detritylation: Removal of the DMTr protecting group.
3. Elongation with the next base, activated with a beta-cyanoethyl group.
4. Capping (irreversible): The non-elongated chains to avoid N − 1 long, wrong chains.
5. Oxidation: Removal of the side group from the added base.
6. Replication (back to point 2) until the length N is reached.

After the synthesis, the following steps are carried out:

1. Cleavage from substrate and collection of product.
2. Removal of the protecting groups.
3. Purification.
4. Control analysis.

Phosphoramidite synthesis can be performed not only ex situ, for example, in a tube, for later use by pipetting or spotting, but also in situ, that is, directly on the final microarray chip. These two options are discussed in the following sections. More methods are described in the review [17].

Question 5

How is produced the DNA for chip functionalization?

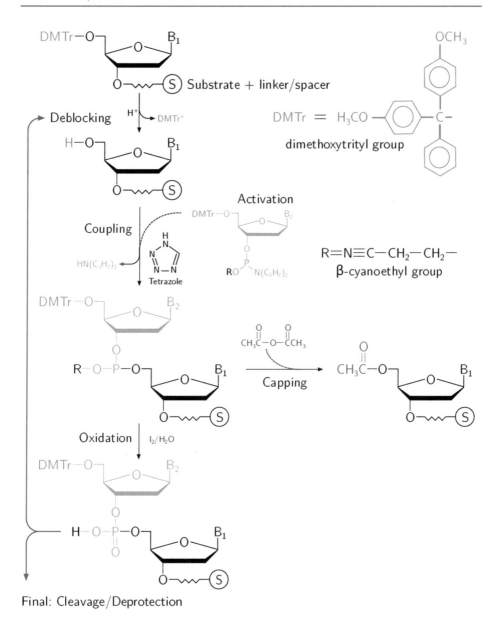

Final: Cleavage/Deprotection

Fig. 8.21 DNA synthesis with phosphoramidite chemistry

8.7.2   Chip Functionalization with Microarray Spotter

After the "ex situ" synthesis (i.e., in a tube and not directly on chip), standard microscope slides can be functionalized with a *microarray spotter* [18, 19]. Every individual DNA probe (sequence) to be spotted must be prepared in advance with the phosphoramidite synthesis discussed above. This requirement limits de facto the applicability of this technique to arrays having at most a few hundreds of probes. It is therefore a low-density arraying method, adequate for routine tasks focusing on a small, well-defined variety of DNA sequences, for instance in the cancer diagnostics based on a set of a few tens of characteristic marker sequences.

The whole *DNA library*, that is, the set of probes foreseen for functionalizing a certain array, is loaded in a contiguous region of a microtiter plate, having the same pitch of the spotting head, called *pintool*, that is, an array of spotting pins. The spot size is typically of 0.1–0.3 mm. Multiple libraries can be loaded in different regions of the same microtiter plate; thus the spotter can functionalize several sets of glass slides, each one with a different library, or even the same slides with multiple subsets of libraries, within one process run. Before moving from one library to another, the pintool is thoroughly washed by dipping it in a second microtiter plate filled with appropriate cleaning solutions (Fig. 8.22).

8.7.3   Lithographic Functionalization: I

The in situ synthesis by lithographic functionalization [21] allows a much higher density of DNA chips. Even though DNA is more stable, also RNA probes can be arrayed conveniently with the same technique.

This proprietary method of Affymetrix [22] consists of the following steps (Fig. 8.23):

1. On a carrier plate (e.g., a quartz wafer), a monolayer of linker molecules (e.g., a SAM) is formed. The linkers are terminated with a photosensitive protective group that prevents coupling reactions with nucleotides.
2. Lithographic masks are used to deprotect specific locations by means of UV light.
3. The surface is flooded with a solution of only one kind of nucleotide: An extension occurs at de-protected locations.
4. The coupled nucleotide bears a light-sensitive protecting group, so the cycle can be repeated.
5. This procedure is repeated for the other three bases, thus completing one layer extension.
6. The entire process is reiterated for further layers until reaching the desired length of normally 25 nucleotides.

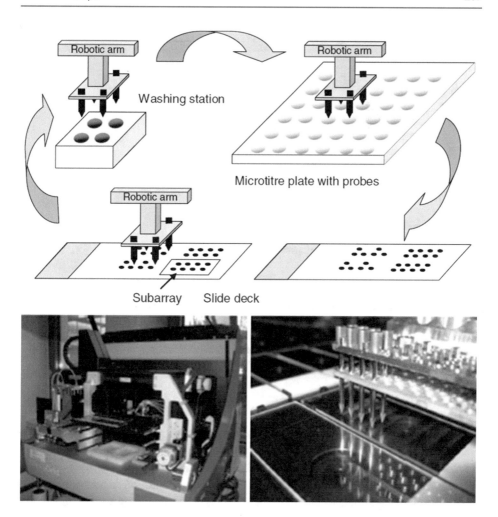

Fig. 8.22 (Top) Working principle of a microarray spotter (Reprinted from [19], © 2009 Humana Press, with permission from Springer Nature). (Bottom, left) BioRobotics MicroGrid II spotting robot and (right) pintool spotting with 8 parallel split pins on epoxysilane-coated slides (Reproduced from [20], © 2017 Springer Science Business Media, with permission from Springer Nature)

With this method, a density of up to 6.5 million spots per array and a spot size of about 5 µm can be achieved. Customization is not required because all possible SNP combinations can be arrayed onto a single chip; however chips with many different functionalization schemes are available, for example, for analyzing the gene expression in cells of different biological species.

The relatively low specificity of the 25 bases' long capturing molecules is compensated by redundancy, that is, multiple spots (e.g., 3) with the same functionalization, and at the software level by means of specific statistical analysis algorithms. Due to the initial investment costs for the set of masks, this method is suitable for large series.

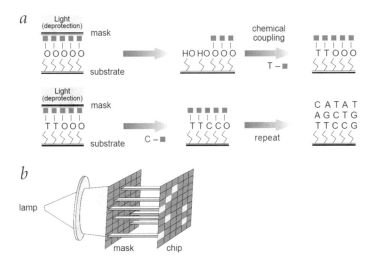

Fig. 8.23 (**a**) Steps of in situ lithographic functionalization: Since each layer usually contains four bases, four lithographic masks for selecting the light patterns of every layer are necessary. The DNA receptors of the Affymetrix microarrays (GeneChips) have a length of 25 bases, therefore a set of 100 masks is required for the whole functionalization procedure. (**b**) Scheme of the projection setup (Reprinted from [21], © 1999, with permission from Springer Nature)

8.7.4 Lithographic Functionalization: II

The high number of masks required for the previous method makes it relatively expensive and affordable only for production of large series. A second method of lithographic functionalization does not require masks. This is the so-called *Maskless Array Synthesizer* (MAS) technology [23], which can be conveniently used for tailor-made high-density chip functionalization. With this technology, the exposure is carried out by a solid-state mirror array, fabricated by means of micro-electromechanical system (MEMS) technologies [24], which is called *Digital Micromirror Device* (DMD) [25, 26].

The density of early devices was of 195,000 spots; later generations of DMDs scaled up the arrays to 786,000 and later to 4.2 million spots. The corresponding miniature aluminum mirrors are individually addressable by software to generate the exposure pattern. Conventional lithographic masks are therefore no longer necessary. The patterns are coordinated with the DNA photo-mediated synthesis chemistry in a parallel, combinatorial manner such that up to 2.1 million distinct probe features, in duplicate, are synthesized on a single array (Fig. 8.24).

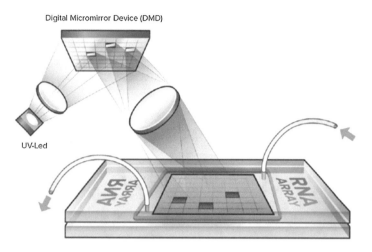

Fig. 8.24 Maskless lithographic functionalization of an RNA-array using DMD. Reagents are supplied through the fluidic cell (Reprinted from [27], © 2018 by the authors, CC-BY 4.0 [8])

8.8   Fluorescence Readout

Laser-induced fluorescence is the most common readout method for microarrays (Fig. 8.25). It fits the whole range from very low to very high density. Both sensitivity and dynamic range are good, and the carrier plates do not need any active component or complex technological fabrication processes.

We have already discussed the fundamentals of fluorescence in Chap. 4. In principle, they apply also for microarrays: A laser scanner examines the arrays and fluorescence light is measured. In case of DNA chips and protein microarrays (see later), which are often used in comparative protocols, the scanners have multiple light sources with different colors, for example, three lasers emitting in the visible range in red, green, and blue available for distinguishing different markers, which fluoresce in different spectral regions. For DNA chips, the typical scanning time required for reading-out one slide is of a few minutes.

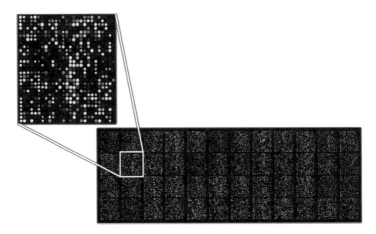

Fig. 8.25 Fluorescence image of a medium-density DNA chip having 12×4 regions with 27×29 spots each (Adapted from [28], CC0 1.0 [8])

8.9   Other (Label-Free) Readout Methods

Similar to what was observed in Chaps. 4 and 5 in case of affinity sensors, the fluorescence readout delivers in most cases a static one-time image of the results. Furthermore, in those cases in which a relatively small number of sequences is investigated, a microarray scanner may be an over-dimensioned instrument. Other techniques address the above two issues, allowing dynamic detections able to deliver information on the kinetics of coupling, by using in some cases a simpler scalable and even portable readout instrumentation.

Among several approaches, it is worth mentioning, Nanowire Field Effect Transistor (Nano-FET) arrays [29], SPR Imaging [30], QCM-arrays [31, 32], Electrode Arrays for the electrochemical redox detection [33–36] and Electrochemical Impedance Spectroscopy [37]. The principles of operation of these approaches are illustrated in the following panel (Fig. 8.26):

Question 6

What are the benefits of lithographic MAS technology compared to mask-based lithographic functionalization?

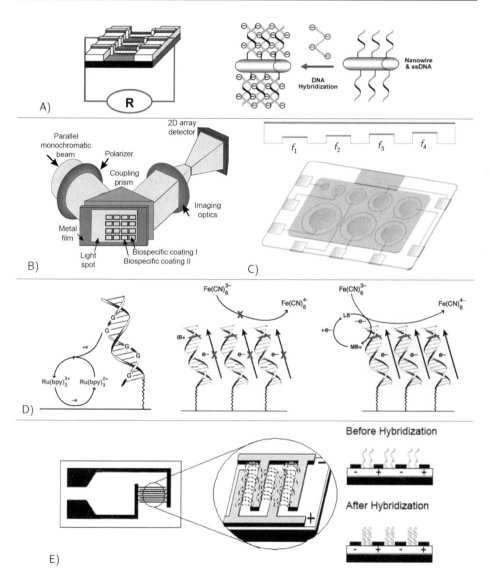

Fig. 8.26 (**a**) Nanowire FET array: Hybridization modulates the drain current (Adapted from [29], © 2007, with permission from the American Chemical Society). (**b**) SPR imaging: Based on changes of reflectivity on each individual spot, uses a 2D image sensor (Reprinted from [30], © 2005, with permission from Elsevier). (**c**) Monolithic multichannel QCM: Individual size and thinning provides unique resonating frequencies for each resonating spot (Reprinted from [32], © 2011, with permission from Elsevier). (**d**) Electrochemical detection with (left) guanine oxidation mediated by a ruthenium complex in solution, and (right) discrimination of mismatches through DNA-mediated charge transport (Adapted from [33], © 2003, with permission from Springer Nature). (**e**) Electrochemical Impedance Spectroscopy with interdigitated electrodes (Reprinted from [37], © 2004, with permission from Elsevier)

8.10 Cancer Diagnostics

In the last part of this chapter, we analyze the procedure for the cancer diagnosis carried out by performing the gene-expression analysis with DNA chips.

In Fig. 8.27, the gene expression of a healthy reference sample is compared with the gene expression of a tissue sample (test) taken from the same organ of a patient under suspicion of cancer. The cDNA of the healthy sample is labeled with fluorophore Cy5 emitting in red, while the test sample is labeled with fluorophore Cy3 emitting in green. The slide is functionalized with a library of tumor-specific prognostic marker genes. After the fluorescence detection, the red spots show genes expressed only by the healthy tissue. The green spots, on the other hand, show genes expressed only by pathologic cells. The spots with mixed colors (light green, yellow, and orange) show genes that are expressed in different ratios by both samples.

8.10.1 Analysis Software

After the fluorescence detection, the collected information is evaluated with the aid of analysis software. A typical data analysis workflow for gene-expression studies based on DNA chips is illustrated in Fig. 8.28. Raw data preprocessing includes steps like log transformation (from linear to logarithmic scale), normalization, and quality control and

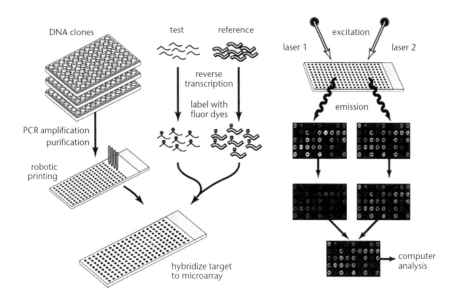

Fig. 8.27 Workflow example for cancer diagnosis with a cDNA microarray (Reprinted from [38], © 1999, with permission from Springer Nature)

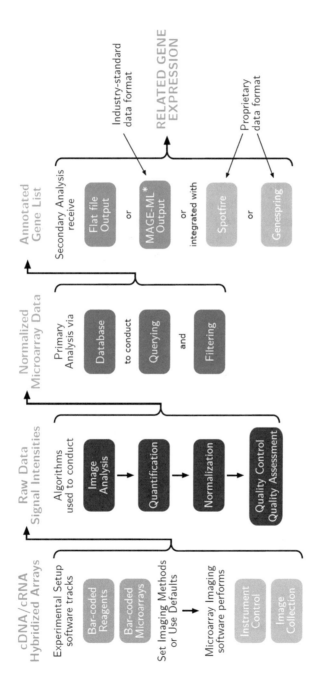

Fig. 8.28 Example of workflow from experimental setup to annotated gene lists, for the data of cDNA microarrays (*MAGE-ML: microarray gene expression markup language)

Table 8.2 Example of		R (intensity of red)	G (intensity of green)
fluorescent signal intensities detected on the spots of a DNA	Spot 1	1251.8	1523.6
chip in a comparative gene	Spot 2	94,326.1	27,875.4
expression analysis (G: patient,	Spot 3	891.6	185.3
R: control reference)	Spot 4	......	......

assessment, which includes handling of replicates and missing values, pattern filtering, and standardization [39].

During the secondary analysis, the data is supplemented through the usage of a fully searchable database, which contains annotations for the existing sequences on the microarray from the publicly accessible database. These annotations contain gene names, acronyms, and references to searchable fields. The primary DNA database is the International Nucleotide Sequence Database Collaboration (INSDC [40]), contributed by the DNA Data Bank of Japan (DDBJ), the European Bioinformatics Institute (EMBL), and the National Center for Biotechnology Information (Genbank). Many other secondary and specialized databanks exist, like the Gene Expression Omnibus (GEO [41]), some of them are public and/or freely accessible and others are commercial. Most annotation outputs are based on flat ASCII or standard spreadsheets formats; however, there are also proprietary formats related to commercial software packages like the GeneSpring by Agilent [42] and the Spotfire by TIBCO [43]; the latter is based on artificial intelligence (AI).

The software analysis starts with the digital image raw data, consisting of an array of numbers (one column per color), which represents the detected red and green intensities for every spot (Table 8.2).

In the next analysis step, data are scale transformed as $\log_2$ of the ratio of the two color intensities, since it is symmetric around ratio = 1 (no change with respect to the reference) (Fig. 8.29):

$$\log \text{ratio}_i = \log_2 \frac{R_i}{G_i}$$

8.10.2 Normalization

Normalization is generally necessary to correct systematic differences between samples on the same slides, or between slides that do not represent true biological variation between the sample (G, patient) with respect to the control reference (R, normal). Actually, the main step is to compensate different fluorescence intensities of the Red (Cy5) and Green (Cy3) dyes.

The normalization can take place in different ways. The two most important are as follows:

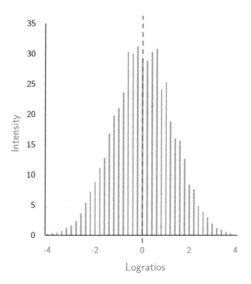

Fig. 8.29 A distribution of logratios is expected to be symmetric around 0

1. *Global normalization (one correction coefficient for all spots)*:

    This is a commonly used method, for example, in the GenePix microarray systems by Molecular Devices, in those cases when the red/green ratio is expected to be constant over the whole array:

$$R_i = k \cdot G_i$$

2. *LOcally WEighted Scatterplot Smoothing (LOWESS)*:

    This intensity-depended normalization corrects the red/green ratio individually for each spot, thus addressing nonlinearities and/or nonuniform fluorescence in different regions of the array. In practice a *scatterplot* is generated placing A_i, the average of the logarithmic intensities of red and green for each spot on the x axis, and M_i, the logratios on the y axis.

$$A_i = \frac{\log_2 R_i + \log_2 G_i}{2} = \log_2 \sqrt{R_i \cdot G_i} \qquad M_i = \log_2 R_i - \log_2 G_i = \log_2 \frac{R_i}{G_i}$$

The LOWESS normalization [44] adjusts the average of the M_i values to the expected zero by providing for every (A_i, M_i) pair a local normalization coefficient $c_i = c(A_i)$. As a result, the centerline of the *MA* scatterplot is corrected (Fig. 8.30). The correction factors c_i are calculated by means of a nonlinear locally weighted polynomial regression of the intensity scatterplot.

$$M_{i,\text{corrected}} = \log_2 \frac{R_i}{G_i} - c(A_i) = \log_2 \frac{R_i}{k_i \cdot G_i} \qquad with: \quad c(A_i) = \log_2(k_i)$$

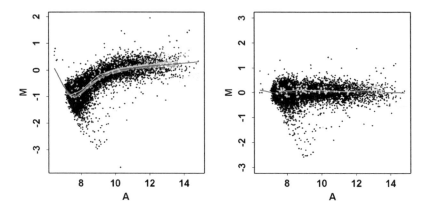

Fig. 8.30 Example of LOWESS normalization: (Left) MA scatterplot before normalization, (right) after normalization. Yellow, green, and orange dots highlight control genes (Reprinted form [44], © 2002, Oxford University Press, CC-BY [8])

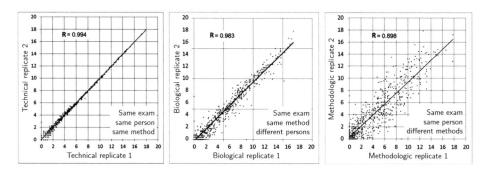

Fig. 8.31 Examples of R for technical (left), biological (middle), and methodologic (right) replicates. Reduced correlation when comparing biological replicates suggests individual differences in gene expression. Reduced correlation between methodological replicates indicates different sensitivities by detecting certain genes (Adapted from [46], © 2013 by the authors, licensee BioMed Central Ltd., CC-BY 2.0 [8])

8.10.3 Pearson's R

The Pearson's correlation coefficient R [45] is a parameter describing data reproducibility. High value of R, that is, close to 1, results when comparing data from two microarrays hybridized with cDNA from the same tissue (technical replicates). By contrast, a reduced R results when comparing the expression levels of biological replicates (i.e., from tissues with the same disease but from different patients). This suggests individual differences in gene expression (Fig. 8.31).

Why is a normalization necessary after the fluorescence readout?
Which methods do you know?

8.11 Clustering

When the same exam is carried out with different patients, we observe biological variations due to differences in the individual expression of the same set of genes. Such differences are relevant not only for distinguishing between healthy and nonhealthy tissues, but allow at the same time a fine differentiation of the diagnosis, which can lead to a personalized treatment plan.

In order to better take into account and analyze the variation of gene expression from different patients, their data are divided into groups based on certain criteria of similarity. This approach is called *Cluster analysis* or simply Clustering and comprises many different algorithms able to extract, even though with different levels of correctness and efficiency, relevant diagnostic and prognostic information from the expression profiles and to visualize the data. Based on mathematical parameters like distance, density, mean, model, etc., or properties like agglomerative, divisive, average, etc., more than 100 clustering algorithms, broadly classified as partitioning or hierarchical approaches, have been published so far. However, clustering is not an objective procedure; its suitability for a given task must be carefully evaluated experimentally, which makes the choice of the right algorithm a challenging undertaking. A thorough description of theory and applications of cluster analysis can be found in [47].

After appropriate preprocessing steps, the datasets (i.e., normalized R–G logratios) are ordered in a matrix, one row per gene, one column per patient (Fig. 8.32) and then the clustering procedure starts. The output results, that is, the discovered groups of data with

Same pathology, different patients

geneID	slide1	slide2	slide3	slide4	slide5	slide6	slide7	...
AB017136	0.48	0.42	0.56	0.22	-0.35	0.1	-0.17	...
AB031049	0.71	0.76	0.55	0.92	0.16	0.2	0.3	...
AF026537	0.54	0.15	0.26	0.44	-0.34	-0.01	-0.32	...
AF071068	NA	0.03	0.04	0.15	-0.24	-0.14	-0.33	...
AF247132	0.41	0.7	0.61	0.55	0.15	-0.24	-0.14	...
...	...	...	...	...	...	...	...	...

values of $\log_2(\text{Red}/\text{Green})$

Fig. 8.32 Biological variations in different patients

similar expression levels, are plotted in so-called *heatmaps* whose color scale corresponds to the whole range of logratio values, for example, from −3 to +3.

In the following, we look closer at two clustering methods, which are among the most commonly adopted ones for gene expression analysis, namely:

1. Clustering by *Density* or *Euclidean distance*: Produces clusters by similar straight-line distance between points.
2. *Hierarchical* clustering: Produces sequences of superclusters (parents) and nested subclusters (children) organized in a tree.

The principle of operation of these algorithms is illustrated in the 2D plots of Fig. 8.33, with two datasets.

In these 2D plots, each point represents a gene plotted against its expression value (logratio) under two experimental conditions (e.g., different samples or different patients with the same disease). The results are represented in the two-columns heatmaps at the right of each plot, with the corresponding cluster identities given in four colors.

These are just two examples of possible clustering strategies. Other methods which work well with gene expression analysis are the Self-Organizing Map (SOM) and the Generative Topographic Map (GTM). They form clusters with the help of learning machines based on artificial neural networks [48]. Comprehensive descriptions of cluster analysis for microarrays and gene expression are given in [49, 50].

The gene expression profile can predict the outcome of a disease and provide indications for patient-tailored therapy strategies. A broad statistical analysis can be used for instance for classifying breast tumors in groups with positive and negative prognoses [51] as depicted in Fig. 8.34.

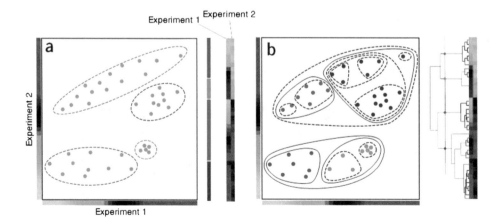

Fig. 8.33 Two strategies for clustering with 40 genes, density (left) and hierarchical (right). These 2D plots represent a scenario with two datasets (e.g., two patients) (Adapted by permission from [48], Springer Nature © 2005)

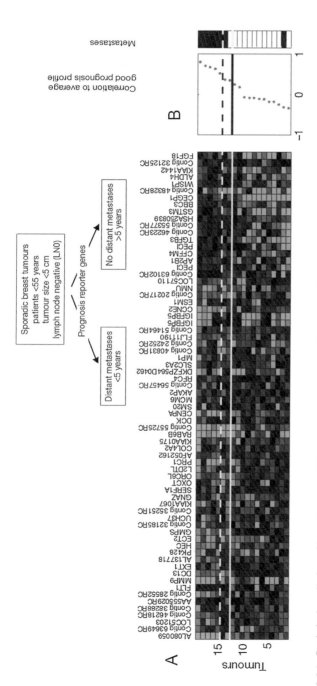

Fig. 8.34 Statistical prediction of breast tumor: (**a**) Expression data matrix of 70 prognostic marker genes from tumors of 19 breast cancer patients. Each row represents a tumor and each column a gene (ordered by correlation with the prognostic groups). Solid yellow line: prognostic classifier with optimal accuracy; dashed line, with optimized sensitivity. Above the dashed line, patients have a good prognosis signature; below the line, the prognosis signature is poor. (**b**) White markings represent patients who developed distant metastases; in black no metastases formation was detected for at least 5 years (Adapted by permission from [51], Springer Nature © 2002)

Question 8

What is the purpose of clustering? Give two examples of clustering strategies!

Answer 1

- A biochip consists of arrayed sensors, therefore multiple analytes can be detected in the same sample, in one run.
- Biochips are miniaturized, and individual sensors are at the microscale or even at the nanoscale; therefore, the required sample quantities are minimal.
- For the same reason, the consumption of scarce and expensive reagents is low.
- Miniaturization may allow portability, low energy consumption, etc.
- Measurements are fast and highly parallelized, yielding a high throughput.

Answer 2

A possesses only two polar sites on the hexagonal ring, one N and one NH_2; therefore, it cannot build a third H bond. T and G cannot pair because their respective polar sites are not complementary, T has O, NH, O, while G has O, NH, NH_2, thus only the respective third site O and NH_2 could form an H-bond, the other two, namely O with O and NH with NH cannot bind.

Answer 3

It is the messenger RNA, which results at the end of transcription after the splicing. It is called "messenger" because mRNA by leaving the cell nucleus transports the instructions for the synthesis of the peptide chain in the translation step.

Answer 4

The PCR is a cyclic (chain) process for amplifying DNA fragments. Amplification means make multiple copies of those molecules. Every cycle consists of three steps at different temperatures: (1) Denaturation at ~95 °C, splits the double-stranded DNA. (2) Primer annealing at ~55 °C for fitting the starting structures. (3) Extension at ~72 °C bases added one by one with the help of the enzyme TAQ-polymerase. The cycle is repeated n times, yielding an amplification of $2n$.

The PCR is needed when the amount of DNA in the available sample is too low for an accurate analysis.

Answer 5

By beta-cyanoethyl phosphoramidite synthesis. This process can be run ex situ, that is, in tubes, for later use, for example, by spotting low-density arrays, or in situ, that is, for the direct functionalization of high-density arrays.

Answer 6

MAS allows custom lithographic functionalization of high-density DNA chips without the expense of the set of masks, needed in case of the Affymetrix method based on standard lithography.

Answer 7

Red and green fluorescent labels have different yields, that is, efficiency by the excitation/emission process. Moreover, some systems may show nonlinearities different sensitives in different regions of a given array. Normalization addresses these issues.

Global normalization provides one correction factor for the whole chip, addressing only the differences of the two labels. *LOWESS* provides a tailored "locally weighted" correction able to address the other problems.

Answer 8

Clustering is the common scope of a wide family of algorithms, designed for grouping data points on the basis of certain similarities, which are defined in terms of mathematical parameters like distance, density, mean, etc. and/or properties like agglomerative, divisive, etc. The discovered groups are visualized in plots called *heatmaps*.

In this chapter, we had a look at the *Euclidean distance or Density* clustering the *Hierarchical* clustering.

Take-Home Messages
- Biochips are a special class of biosensors; they are miniaturized and consist of integrated arrays of sensors, ranging from a few up to millions of sensing units per chip.
- This approach allows high sample throughput and parallelization of analysis with low consumption of samples and reagents.
- The most important application of biochips is the analysis of gene expression with DNA microarrays often called DNA chips.
- Captured mRNA form cells is converted to cDNA by reverse transcriptase, then amplified, controlled, and analyzed.
- DNA-chip are functionalized with known DNA sequences (oligonucleotides) produced by design with the beta-cyanoethyl phosphoramidite synthesis.
- A specific set of functionalizing DNA sequences used for a specific analysis is called *gene library* and collects the marker genes, which are relevant for that purpose, for example, a cancer diagnosis or the determine the effects of a drug under development.

(continued)

- After the hybridization and wash steps, the DNA arrays are inspected in most cases by fluorescence, however also label-free techniques are available. Here we mentioned nano-FET arrays, SPR imaging, QCM arrays, MEAs, and EIS.
- Fluorescence readout data are first preprocessed, that is, log-transformed, normalized, controlled for quality, and corrected as needed. Then database queries lead to an annotated gene list as output.
- Data from comparative clinical studies on groups of patients are conveniently processed by clustering, in order to find similarities and dissimilarities with diagnostic and prognostic relevance, which lead to a carefully tailored therapy (i.e., personalized medicine).

References

1. Van Norman GA. Drugs, devices, and the FDA: part 1: an overview of approval processes for drugs. JACC. 2016;1(3):170–9.
2. Wecker L, Krzanowski J. The drug approval process in Japan. In: Reference module in biomedical sciences. Amsterdam: Elsevier; 2014.
3. Van Norman GA. Drugs and devices: comparison of European and U.S. approval processes. JACC. 2016;1(5):399–412.
4. Wecker L, Krzanowski J. The drug approval process in Europe. In: Reference module in biomedical sciences. Amsterdam: Elsevier; 2014.
5. Yao X, Ding J, Liu Y, Li P. The new drug conditional approval process in China: challenges and opportunities. Clin Ther. 2017;39(5):1040–51.
6. Bano E, Fradetal L, Ollivier M, Choi J-H, Stambouli V. SiC nanowire-based transistors for electrical DNA detection. In: Saddow SE, editor. Silicon carbide biotechnology. 2nd ed. Elsevier; 2016. p. 261–310.
7. Smedlib. Gene Intron Exon nb. https://commons.wikimedia.org/wiki/File:Gene_Intron_Exon_nb.svg.
8. The Creative Commons copyright licenses. https://creativecommons.org/licenses/.
9. Roberts RJ, Vincze T, Posfai J, Macelis D. REBASE--a database for DNA restriction and modification: enzymes, genes and genomes. Nucleic Acids Res. 2015;43:D298–9.
10. REBASE—the restriction enzyme database. http://rebase.neb.com/rebase/rebase.html.
11. Saiki RK, Scharf S, Faloona F, Mullis KB, Horn GT, Erlich HA, et al. Enzymatic amplification of beta-globin genomic sequences and restriction site analysis for diagnosis of sickle cell anemia. Science. 1985;230(4732):1350–4.
12. Enzoklop. Schematic mechanism of PCR.
13. Chou J, Wong J, Christodoulides N, Floriano PN, Sanchez X, McDevitt J. Porous bead-based diagnostic platforms: bridging the gaps in healthcare. Sensors. 2012;12(11):15467–99.
14. van Pelt-Verkuil E, van Belkum A, Hays JP. Important considerations for typical, quantitative and real-time pcr protocols. In: Principles and technical aspects of PCR amplification. Dordrecht: Springer; 2008. p. 119–39.
15. van Pelt-Verkuil E, van Belkum A, Hays JP. Ensuring PCR quality – laboratory organisation, PCR optimization and controls. In: Principles and technical aspects of PCR amplification. Dordrecht: Springer; 2008. p. 183–212.

16. Caruthers MH, Beaucage SL, Becker C, Efcavitch JW, Fisher EF, Galluppi G, et al. Deoxyoligonucleotide synthesis via the phosphoramidite method. Gene Amplif Anal. 1983;3:1–26.

17. Miller MB, Tang Y-W. Basic concepts of microarrays and potential applications in clinical microbiology. Clin Microbiol Rev. 2009;22(4):611.

18. DeRisi J, Penland L, Brown PO, Bittner ML, Meltzer PS, Ray M, et al. Use of a cDNA microarray to analyse gene expression patterns in human cancer. Nat Genet. 1996;14(4):457–60.

19. Dufva M. Fabrication of DNA microarray. In: Dufva M, editor. DNA microarrays for biomedical research: methods and protocols. Totowa: Humana Press; 2009. p. 63–79.

20. Rupp S. Microarray technologies in fungal diagnostics. In: Lion T, editor. Human fungal pathogen identification: methods and protocols. New York: Springer; 2017. p. 385–409.

21. Lipshutz RJ, Fodor SPA, Gingeras TR, Lockhart DJ. High density synthetic oligonucleotide arrays. Nat Genet. 1999;21(1):20–4.

22. Affymetrix webpage. https://www.thermofisher.com/de/de/home/life-science/microarray-analy sis/affymetrix.html?navMode=35810&aId=productsNav.

23. Nuwaysir EF, Huang W, Albert TJ, Singh J, Nuwaysir K, Pitas A, et al. Gene expression analysis using oligonucleotide arrays produced by maskless photolithography. Genome Res. 2002;12 (11):1749–55.

24. Huang QA. Micro electro mechanical systems. Singapore: Springer; 2018. p. 1479.

25. Monk DW, Gale RO. The digital micromirror device for projection display. Microelectron Eng. 1995;27(1):489–93.

26. Song Y, Panas RM, Hopkins JB. A review of micromirror arrays. Precis Eng. 2018;51:729–61.

27. Lietard J, Ameur D, Damha MJ, Somoza MM. High-density RNA microarrays synthesized in situ by photolithography. Angew Chem Int Ed. 2018;57(46):15257–61.

28. Lu H, Giordano F, Ning Z. Oxford nanopore MinION sequencing and genome assembly. Genomics Proteomics Bioinformatics. 2016;14(5):265–79.

29. Gao Z, Agarwal A, Trigg AD, Singh N, Fang C, Tung C-H, et al. Silicon nanowire arrays for label-free detection of DNA. Anal Chem. 2007;79(9):3291–7.

30. Homola J, Vaisocherová H, Dostálek J, Piliarik M. Multi-analyte surface plasmon resonance biosensing. Methods. 2005;37(1):26–36.

31. Abe T, Esashi M. One-chip multichannel quartz crystal microbalance (QCM) fabricated by deep RIE. Sensors Actuators A Phys. 2000;82(1):139–43.

32. Tuantranont A, Wisitsora-at A, Sritongkham P, Jaruwongrungsee K. A review of monolithic multichannel quartz crystal microbalance: a review. Anal Chim Acta. 2011;687(2):114–28.

33. Drummond TG, Hill MG, Barton JK. Electrochemical DNA sensors. Nat Biotechnol. 2003;21 (10):1192–9.

34. Furst A, Landefeld S, Hill MG, Barton JK. Electrochemical patterning and detection of DNA arrays on a two-electrode platform. J Am Chem Soc. 2013;135(51):19099–102.

35. Schienle M, Paulus C, Frey A, Hofmann F, Holzapfl B, Schindler-Bauer P, et al. A fully electronic DNA sensor with 128 positions and in-pixel a/D conversion. IEEE J Solid State Circuits. 2004;39(12):2438–45.

36. Dodel N, Laifi A, Eberhardt D, Keil S, Sturm L, Klumpp A, et al., eds. A CMOS-based electrochemical DNA hybridization microarray for diagnostic applications with 109 test sites. In: IEEE biomedical circuits and systems conference (BioCAS). 2019; 17–19 Oct. 2019.

37. Hang TC, Guiseppi-Elie A. Frequency dependent and surface characterization of DNA immobilization and hybridization. Biosens Bioelectron. 2004;19(11):1537–48.

38. Duggan DJ, Bittner M, Chen Y, Meltzer P, Trent JM. Expression profiling using cDNA microarrays. Nat Genet. 1999;21(1):10–4.

39. Herrero J, Díaz-Uriarte R, Dopazo J. Gene expression data preprocessing. Bioinformatics. 2003;19(5):655–6.
40. International Nucleotide Sequence Database Collaboration. http://www.insdc.org/.
41. Mardis EM. Next-generation sequencing technologies. 2016.
42. Genespring by Agilent. http://genespring-support.com/.
43. Spotfire by TIBCO. https://www.tibco.com/products/tibco-spotfire.
44. Yang YH, Dudoit S, Luu P, Lin DM, Peng V, Ngai J, et al. Normalization for cDNA microarray data: a robust composite method addressing single and multiple slide systematic variation. Nucleic Acids Res. 2002;30(4):15.
45. Jiong Y, Wei W, Haixun W, Yu P. /spl delta/−clusters: capturing subspace correlation in a large data set. In: Proceedings 18th international conference on data engineering. 2002; 26 Feb–1 Mar 2002.
46. Huang X, Yuan T, Tschannen M, Sun Z, Jacob H, Du M, et al. Characterization of human plasma-derived exosomal RNAs by deep sequencing. BMC Genomics. 2013;14(1):319.
47. Gan G, Ma C, Wu J. Data clustering: theory, algorithms, and applications. 2nd ed. Philadelphia: Society for Industrial and Applied Mathematics; 2020. p. 431.
48. D'Haeseleer P. How does gene expression clustering work? Nat Biotechnol. 2005;23 (12):1499–501.
49. Franco M, Vivo J-M. Cluster analysis of microarray data. In: Bolón-Canedo V, Alonso-Betanzos-A, editors. Microarray bioinformatics. New York: Springer; 2019. p. 153–83.
50. Gan G, Ma C, Wu J. Clustering gene expression data. In: Data clustering: theory, algorithms, and applications. 2nd ed. Philadelphia: Society for Industrial and Applied Mathematics; 2020. p. 329–40.
51. van Veer LJ, Dai H, van de Vijver MJ, He YD, AAM H, Mao M, et al. Gene expression profiling predicts clinical outcome of breast cancer. Nature. 2002;415(6871):530–6.

Genome Sequencing

9

Contents

Keywords

Dideoxynucleotide · Electropherogram · Bridge amplification · Emulsion PCR ·
Nanopores · Genome assembly · Human genome reference

© The Author(s), under exclusive license to Springer Nature Switzerland AG 2021   225
A. Pasquarelli, *Biosensors and Biochips*, Learning Materials in Biosciences,
https://doi.org/10.1007/978-3-030-76469-2_9

What You Will Learn in This Chapter

In this chapter, you will learn how to read genomic DNA with Sanger method and Next-Generation Sequencing techniques. We will cover natural as well as synthetic nanopores and their respective properties. We also study how the nucleotide sequences can be scanned as they pass through such pores.

You will be able to explain the functional principle of Massively Parallel DNA Sequencing devices as well as how nucleotide identification works based on the current intensity profile while passing DNA strands through suitable nanopores.

In the previous chapter, we learned how to monitor the proteinogenic cell activity by analyzing the gene expression with DNA chips. In this chapter, we expand our analysis capability on the whole genome, for detecting also the non-expressed genes and the regulatory sequences. This is obtained with *genome sequencing* methods, which allow "to read" the whole DNA stored in a cell, that is called *genomic DNA*, short *gDNA*.

The sequencing techniques are classified in terms of their historical development as follows:

- First Generation: *Sanger method*, based on restriction enzymes, PCR, Gel electrophoresis.
- Second-Generation Sequencing (NGS): Sequencing-by-synthesis, after amplification.
- Third Generation: Single molecule sequencing, no amplification needed.

9.1   First-Generation DNA Sequencing (Sanger Method)

DNA sequencing is the determination of the nucleotide sequence in a DNA molecule. There are several methods for sequencing DNA. The Sanger method [1], developed in 1977, and considered the gold-standard for more than 25 years, uses the polymerase chain reaction (PCR) to produce single-stranded DNA molecules, differing in length by one single nucleotide, that can be separated using gel electrophoresis. The inclusion of a small amount of *dideoxynucleotide* bases in the mixture of all four normal bases causes chain termination during PCR. The probabilistic binding of a dideoxynucleotide triphosphate (ddNTP) leads to chain termination whenever it is incorporated into the extending DNA (Figs. 9.1 and 9.2).

It is important to stress that the inclusion of, for example, *dideoxyadenosine* triphosphate (ddATP) is random and therefore all possible chains that end with "A" will exist. The four different kinds of ddNTPs bear a fluorescent dye with a base-specific color.

The Sanger method is historically very important because it allowed for the first time to sequence whole genomes. Nevertheless, it has a throughput rate of at most a few tens of

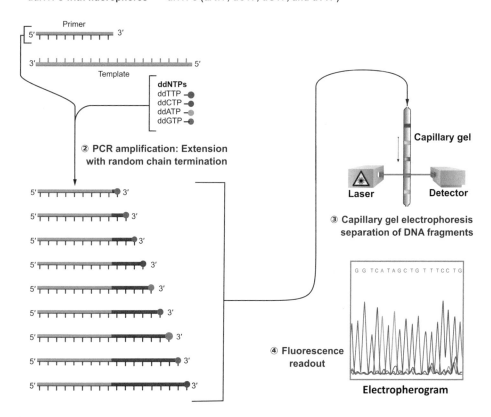

Fig. 9.1 Deoxyribonucleotide triphosphate (dNTP, left) and Dideoxyribonucleotide triphosphate (ddNTP, right). The missing OH group at position 3′ prevents a further extension of the nucleotide chain, once a ddNTP has attached. Yet the number of ddNTPs is relatively low, and therefore all extension lengths have a similar probability

Fig. 9.2 Workflow of Sanger method. (1) Reagents setup. (2) PCR amplification: ddNTPs stop extension randomly, yielding fragments of all possible lengths. (3) Separation of PCR products in a capillary (single lane) gel electrophoresis. (4) Fluorescence readout and data merging into an *electropherogram* (Adapted from [2], © Estevezj, CC-BY-SA 3.0 [3])

kilobases per run (remember a whole genome is in the range of gigabases). In the early 2000s new technologies were developed, which quickly boosted the throughput rate by several orders of magnitude (Fig. 9.3). These new technologies are called *Next-Generation Sequencing* (NGS).

Question 1

What is the definition of DNA sequencing?

Which steps are necessary according to the Sanger method?

9.2   Next-Generation Sequencing (Second Generation)

Next-Generation Sequencing (NGS) is based on *Massively Parallel DNA Sequencing (MPS)* instruments. All MPS platforms require a DNA library obtained by fragmentation of genomic, that is, chromosomal DNA, and ligation with custom linkers (adapters). Each library fragment is amplified on a solid surface (either a bead or a flat surface integrated in the fluidics) with covalently attached adapters that hybridize the library adapters. The direct step-by-step detection of the nucleotide base is incorporated by each amplified library fragment set. Thus, it is possible to detect in parallel hundreds of thousands to hundreds of millions of reactions per instrument run.

The raw DNA obtained from the cells under study needs first to be converted to a library before sequencing can be carried out [5, 6]. Such library preparation consists of the following steps (Fig. 9.4):

- Fragment high molecular weight DNA with sonication
- Enzymatic base-filling, to blunt ends (ends repair)
- Ligate synthetic DNA adapters (carrying a *DNA barcode*)
- Size selection by gel electrophoresis
- PCR amplification
- Proceed to Whole-Genome Sequencing (WGS)

In the following, we give a closer look at two methods belonging to the second generation, namely the Illumina/Solexa approach with fluorescence readout, and the Ion Torrent approach, with electrochemical readout.

9.2.1   The Illumina/Solexa Approach

In this method [8], a glass chip is the carrier on which first the amplification and then the MPS are carried out. The adopted *solid-phase amplification* is a variation of the PCR,

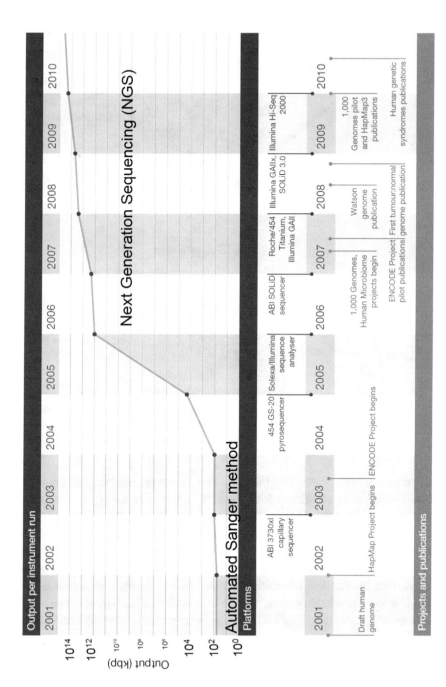

Fig. 9.3 (Top) Progress in instrument sequencing throughput represented on a logarithmic scale. Bottom) Milestones of NGS platforms and sequencing projects (Adapted from [4], © 2011, with permission from Springer Nature)

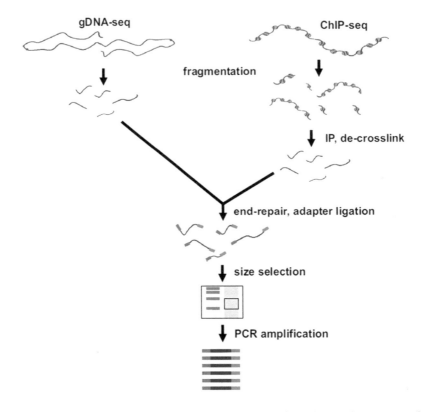

gDNA-seq   ChIP-seq

fragmentation

IP, de-crosslink

end-repair, adapter ligation

size selection

PCR amplification

Fig. 9.4 Simplified scheme showing the library preparation for NGS. (Left) Fragmentation of genomic DNA, or (right) fragmentation after chromatin immunoprecipitation within the ChIP-seq [7] procedure. (Bottom) Common steps including end repair (blunt ending), adapter ligation, size selection, and PCR amplification (Reprinted from [6], © 2014, with permission from Elsevier)

having the primers pre-immobilized on the slide. The strands are replicated several times by *bridge amplification*. Such bridges form spontaneously due to the physical attraction forces related to the primer hybridization. In this way, up to hundreds of millions of local clusters of identical strands are formed in a fashion equivalent to the spots of high-density DNA chips (Fig. 9.5).

Also, the sequencing is cyclic. In the first cycle, the flow cell is flooded with a mixture containing all four labeled and reversibly protected bases, together with primers and polymerase. Due to the protection groups, a single-base extension takes place. Immediately after, fluorescence is detected by laser excitation. In every cluster, several identical labeled bases emit the color corresponding to the incorporated base, thus providing a clearly detectable base-specific color spot, in an image consisting of hundreds of million spots. The image is processed in real time, thus recording the identity of the base for each cluster. Before initiating the next readout cycle, the blocked 3' terminus and the fluorophore from each incorporated base are removed. In the following cycles the procedure is repeated, with

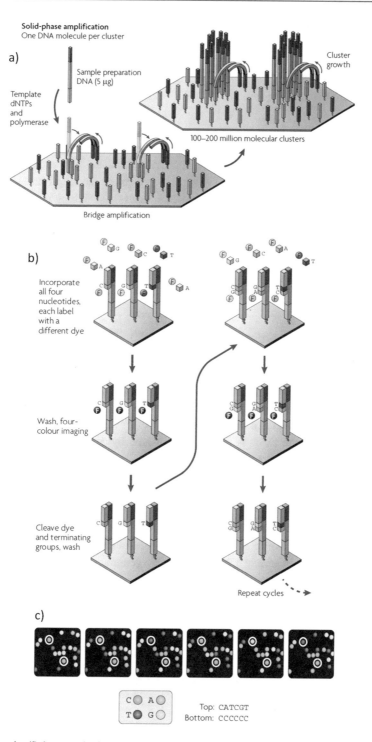

Fig. 9.5 Illumina/Solexa method. (**a**) "Solid-phase amplification" is composed of two basic steps: Initial priming and extension of the single-stranded, single-molecule template, and then bridge

the sole difference that primer molecules are no longer present in the mixture. This methods is suitable for sequencing strands of up to a few hundred bases.

9.2.2   ION Torrent

Another method of the second-generation sequencing is *ION Torrent* [10–12]. Here the detection is performed with a CMOS chip (Fig. 9.6) comprising an array of ion-sensitive field-effect transistor (ISFETS), each one of them surmounted by a micro-well where the sequencing-by-synthesis takes place. The signals detected by the ISFET array is transferred to the output driver by means of a 2D Multiplexer, that is, a switching matrix. This method is based on the fact that chain extension by incorporating a base at the $3'$ position releases a proton (H^+). Thus, the ISFFET underneath the well can detect the related pH transient, which is proportional to the amount of incorporated bases.

ION Torrent chips are available in three capacities of 1, 7, and 12 million wells, named Ion314, Ion316, and Ion318, respectively.

9.2.3   Library Preparation of "DNA Beads"

As seen above, in this platform the sequencing-by-synthesis takes place in the individual wells. Here the DNA strands are not bound to the sensor plate, but on carrier beads to allow an easy cleaning and multiple uses of these hi-tech devices. The size of the beads is tailored to the well size for providing a single-bead capacity per well. Prior to placing the beads in the individual wells the DNA is amplified in a tube by *Emulsion PCR* [13]. The ION Torrent workflow is illustrated in Fig. 9.7.

The library preparation consists of the following steps:

- Get cells sample, isolate DNA and fragment it via sonication.
- Size select with gel electrophoresis (average length 100–160 bases).
- Ligate forward and reverse adapters to the fragments (average length 160–220 bases).
- Emulsify beads [1] containing DNA-oligos with genomic library (diluted such that $\sim$ 1 fragment binds to 1 bead).

Fig. 9.5 (continued) amplification of the immobilized template with immediately adjacent primers to form clusters. (**b**) Parallel fluorescent readout by extending all strands by one base per cycle. (**c**) Sequencing data of the strands at the respective positions are obtained by stacking the images of the detected fluorescence emission (Reprinted from [9], © 2010, with permission from Springer Nature)

---

[1] Dynabeads M-280 Streptavidin (Dynal Corporation, Oslo, Norway).

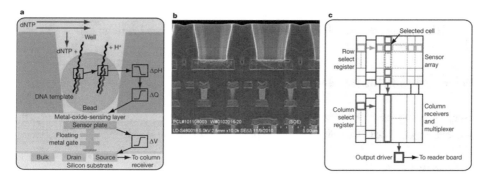

Fig. 9.6 (**a**) ION Torrent principle of operation with one bead per ISFET well. (**b**) SEM picture showing cross-section of the CMOS chip. (**c**) Multiplexer architecture for the signal readout (Reprinted from [10] © 2011, with permission from Springer Nature)

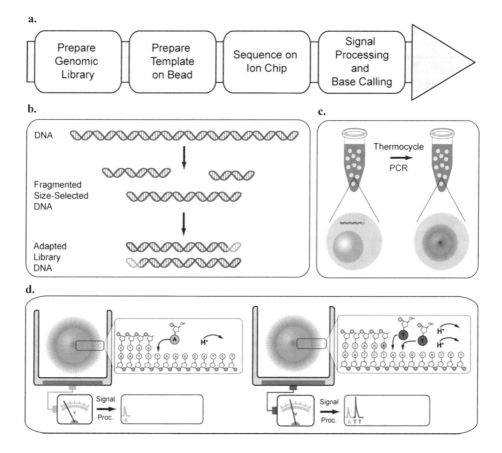

Fig. 9.7 (**a**) Ion Torrent workflow. (**b**) DNA Library preparation. (**c**) Emulsion PCR. (**d**) Electrochemical readout detecting pH transients due to proton release during the reaction. Extension by multiple identical bases is possible in the same cycle and leads to a proportionally larger, for example, twofold, threefold, etc., pH transient (Reprinted from [10] © 2011, with permission from Springer Nature)

- Amplify fragments (on the same bead) with emulsion PCR.
- Release beads and get rid of empty beads with magnetic bead enrichment.

9.2.4   pH Sensing of Base Incorporation

Once the emulsion PCR is ready and the beads carry the desired number of DNA copies the sequencing can start. This procedure comprises the following steps:

- Load beads into chip together with polymerase.
- Load beads into sensor wells by centrifuging (1 bead/well: 2 μm beads, 3.5 μm wells).
- Stepwise DNA hybridization by loading cyclically nucleotides (one by one) with a fluidic system, A, G, C, T, A, G, C, T, etc., repeat (read ALL sensors at every step).
- Feed measurement (each sensor, each step) into software model (taking into consideration diffusion rates, buffering effects, and polymerase rates).
- Exclude low accuracy reads, beads with multiple vectors, beads without DNA.
- Reproduce whole genome by chaining up overlapping sequences.

Question 2
Can you explain the meaning of "sequencing-by-synthesis"?

9.3   Next-Generation Sequencing (Third Generation)

Further developments led to newer methods, which do not need expressly amplification of nucleic acid samples, for example, by PCR. These methods represent the third generation of sequencing. In the following, we give a look at two methods, one based on optical readout, while the other relies on the detection of ionic current fluctuations when DNA stands translocate though suitable nanopores.

9.3.1   Pacific Biosciences

Pacific Biosciences [14] developed biochip systems performing MPS by optical detection and belonging to the third generation. This technology performs sequencing-by-synthesis of single DNA molecules, in other words no PCR amplification is necessary [9, 15, 16].

The sensor plate consists of an array of millions of nano-wells called zero-mode waveguides (ZWMs). Briefly, the ZWM [17, 18] is a hole with a diameter of a few tens of nanometers in a metal film 100 nm thick, deposited on a glass substrate. The volume of a ZMW is ~20 zeptoliters (20×10^{-21} L). When illuminating the carrier at the backside, the light cannot enter the wells, which are much smaller than the wavelength; however, some electromagnetic energy penetrates at a very shallow depth of just a few nanometers. Each ZWM holds one single molecule of the polymerase enzyme immobilized at the bottom (Fig. 9.8 a). Therefore, the extension of DNA takes place within the shallow penetration depth of the light energy.

In this method, the deoxynucleotides carry a base-specific fluorescent label linked at the phosphate side, which is a hexaphosphate group [19] instead of a triphosphate like in all other previously seen methods. Due to the very small volume of a ZMW, the labeled nucleotides have a little probability to be present in the energized layer, if ever for just a few microseconds; therefore the unspecific fluorescence background is very low.

At extension time (Fig. 9.8 b), the complementary base being incorporated is held for several milliseconds within the energized volume, thus permitting the color readout in real time during this step. By forming the phosphodiester bond, the label is cleaved and diffuses in the bulk out of the energized detection volume. Being the label attached at the hexaphosphate, no extra chemistry is necessary for cleavage/deprotection; therefore, the extension can proceed uninterrupted, resulting in a reduced consumption of nucleotides, because solution exchange is not necessary, and a sequencing speed of ~10 bases per second and ZMW.

The emitted light flashes, collected by an objective, pass first through a prism for color separation, resulting in an individual "rainbow," that is, a spectrum for every single ZMW, and then are focused on a monochrome electron-multiplying image sensor (EM-CCD) able to record them with single-photon sensitivity. In this way, a single image sensor can collect at the same time both the spatial information (ZMW coordinates) and the nucleotide-associated information (sequence).

The Pacific Biosciences technology is able to sequence strands of up to a few ten thousand bases. This ability to perform *long reads* is particularly important in the emerging field of medical genetics for the detection of small mutations in rare diseases and the consequent development of specifically tailored therapeutic strategies [20].

Question 3

Can you relate the nucleotide variations shown in Fig. 9.9 to the methods in which they are respectively used?

Compare the polymerase-mediated synthesis with the phosphonamidite synthesis.

9.4 Nanopore Sensing

Nanopore-based DNA sequencing started in the early 1990s with the adoption of α-hemolysin as core device for this sensing approach. After nearly 25 years of development, it became one of the best performing third-generation sequencing methods.

The protein α-hemolysin is a toxin secreted by the round-shaped bacterium *Staphylococcus aureus* and has the following properties:

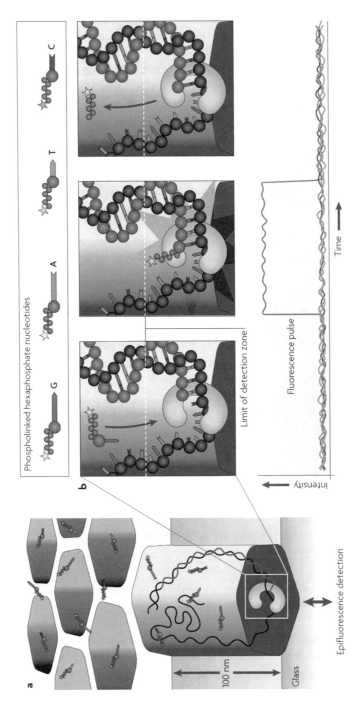

Fig. 9.8 Pacific Biosciences' four-color real-time sequencing method. Polymerase-driven synthesis and light signal: (**a**) Energy from the excitation light beam penetrates only the lower 20–30 nm of each well, creating a detection volume of 20 zeptoliters (10^{-21} L). (**b**) The residence time of phospholinked nucleotides in the active site is on the millisecond scale, which corresponds to the duration of the fluorescence flash (Reprinted from [9]. © 2010, with permission from Springer Nature)

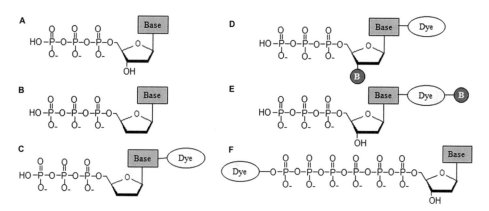

Fig. 9.9 Summary of nucleotide variations used in different polymerase-mediated synthesis methods. (**a**) Deoxynucleotides (dNTPs); (**b**) dideoxynucleotides (ddNTPs); (**c**) dye terminators; (**d**) reversible dye terminators; (**e**) 3′-OH unblocked reversible dye terminators; (**f**) dye-labeled hexaphosphate nucleotides (Adapted from [21], © 2014 by the Authors, CC-BY 3.0 [13])

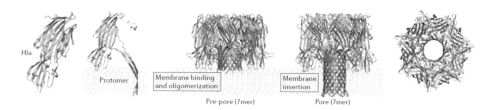

Fig. 9.10 Insertion of the α-hemolysin in an LBM and pore formation (Adapted from [23], © 2016, with permission from Springer Nature)

- Very well-known 3D structure [22].
- Stable over wide ranges of pH and temperature.
- The pore stays open at normal conditions.
- Can bind to various biological or synthetic lipid bilayer membranes (LBMs).
- The binding proceeds spontaneously and does not require specific ionic conditions.

α-hemolysin is therefore a prominent example for the concept of using channels as devices.

Shown below is the insertion of the seven monomeric proteins into the membrane (LBM) and the formation of the heptameric pore. This makes the originally tight LBM permeable to ions (Fig. 9.10).

Due to its robustness, α-hemolysin was considered for long time the ideal nanopore for DNA sequencing. Rather long (kilobases) single-stranded DNA molecules translocate

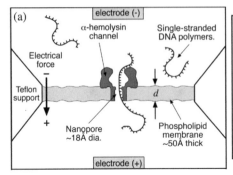
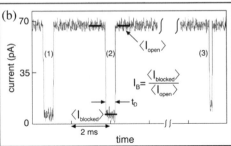

Fig. 9.11 (**a**) Single-stranded DNA molecules are negatively charged; thus, the applied potential induces electrophoretic forces driving DNA together with salt ions through a single α-hemolysin pore inserted in an LBM. During DNA translocation most of the ionic current is blocked. (**b**) Translocation current blockades detected with three polydeoxyadenosine strands poly(dA) of 15 oligonucleotides (1 and 2) and 7 oligonucleotides (3). Each event is characterized by the translocation time, t_D, and the average event blockade, I_B (Reprinted with permission from [24], © 2001, by the American Physical Society)

through the pore when driven by a transmembrane potential, producing ionic current blockades that reflect the structure of individual strands (Fig. 9.11).

Yet it took much more time to establish how single nucleotides can actually be discriminated within the translocating DNA, firstly because the translocation speed of DNA was too fast (∼1 μs per base) to identify individual bases. In 2012 it was proposed to use a DNA polymerase, from the phage phi29, for ratcheting DNA at a rate of about one nucleotide every 10 milliseconds or slower [25]. Such method was suitable for lowering the translocation speed of DNA by four orders of magnitude compared with freely translocating DNA. Secondly, the β-barrel of α-hemolysin resulted too long, with the effect of reducing the sensitivity to single bases.

A first breakthrough was achieved after the proposal of using the phi29 DNA polymerase together with the pore MspA from the *Mycobacterium smegmatis* as a better detector of individual bases in a DNA strand [26]. In fact, MspA has a sensing region of ∼0.6 nm in length, that is, much shorter than the one of α-hemolysin (Fig. 9.12).

The *base-calling* software for the base identification from the current fluctuations was progressing as well. The adoption of the Viterbi algorithm [28] pointed out that the sensing region of the ideal nanopore should be three bases long, thus providing not only nominally different current levels for the 64 different triplets (Fig. 9.13a), but also some data redundancy, being the first two bases present also in the read of the previous triplet and the last two bases present also in the next triplet read (Fig. 9.13b), thus drastically reducing the base-calling errors. This additional knowledge led finally to the adoption of the CsgG pore [29] from *E. coli* (Fig. 9.14) in the chips of Oxford Nanopore Technology, the first commercial nanopore-based sequencing devices.

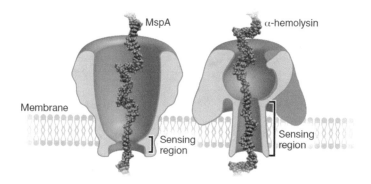

Fig. 9.12 Comparison of the sensing regions in MspA and α-hemolysin (Reprinted from [27], © 2016, with permission from Springer Nature)

9.4.1 Oxford Nanopore Technology (ONT) Sequencing

The ONT biosensing platform [30] performs MPS without amplification and without synthesis (Figs.9.15, and 9.16). It has the unique capability of sequencing very long strands [31], thus easing significantly the "assembly" procedure for reconstructing the genome as a whole (see following section).

The main features of this method are follows:

- Array of 9-meric nanopores assembled in an insulating membrane.
- A polymerase works as a wrench engine yielding a transit time of bases (∼450 bases/s) for optimal speed/quality ratio.
- Each nanopore is controlled and measured by a dedicated Application-Specific Integrated Circuit (ASIC).
- Post-run data analysis compares pore current changes to a model of all possible multimers.
- Variable read lengths, up to 100 kilobases and more.
- Ionic current-based detection of nucleotides in pore, that is, no synthesis reaction needed.

Nanopore sequencing is less accurate than other methods; however, it has some unique features, which provide complementary information hardly achievable with other NGS systems. Since neither amplification nor synthesis are necessary, original nucleic acids can be sequenced thanks to ultra-long reads in a nearly unmodified state, thus easily permitting the detection of large structural variants and epigenetic modifications [31]. Moreover, direct full-length RNA sequencing was reported for the first time using the MinION device [34], opening the way for transcriptome analysis without the biasing risks associated with the reverse transcription and amplification steps.

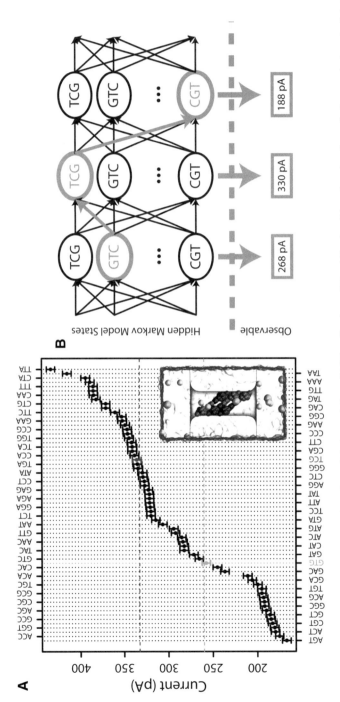

Fig. 9.13 (**a**) Simulation of current levels for all possible base triplets. (**b**) Visualization of Viterbi's algorithm showing the path that maximizes the joint probabilities during base transitions (Reprinted from [28], © 2012, with permission from Elsevier)

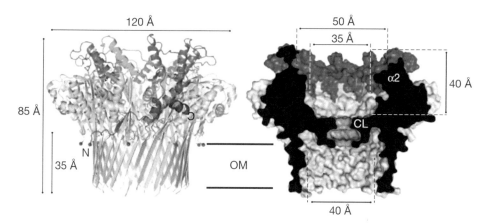

Fig. 9.14 Structure of the CsgG pore from *Escherichia coli*. It is a nonameric (nine subunits) lipoprotein with a 36-strands β-barrel shape. Its dimensions and position of the constriction (CL) make it a better pore for DNA sequencing. For proper use in the Oxford Nanopore Technology devices, the CsgG wild-type has been engineered with more than 700 mutant to enhance its properties (Adapted from [29], © 2014, with permission from Springer Nature)

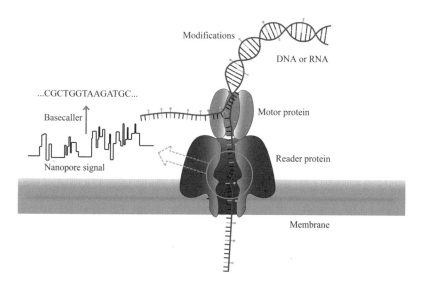

Fig. 9.15 Nanopore sequencing. The polymerase motor protein moves the DNA or RNA strand through the nanopore of the reader protein. Double-stranded molecules are split and only one strand enters the pore. The basecaller converts the current fluctuations detected during the translocation into the corresponding sequence (Reprinted from [32], © 2021, with permission from Springer Nature)

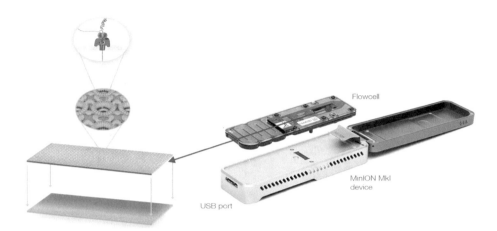

Fig. 9.16 Oxford Nanopore Technology. (Left) Up to 2048 nanopores work in parallel in one chip assembled in a flowcell. (Right) The MinION device is an ultra-compact electronic front-end for acquiring the sequencing signals. This sensing frontend needs just a laptop and an Internet connection for providing a full-featured, high-performance, and portable MPS system (Reprinted from [33], © 2016 by the Authors, hosted by Elsevier, CC-BY 4.0 [3])

Can you summarize in a comparative table the key features of Pacific Bioscience and Oxford Nanopore?

9.5   Genome Assembly

DNA sequencing technologies cannot read whole genomes in one run, but rather read fragments of 20 to 100,000 bases, depending on the adopted technology. *Assembly* is a technique to *align* and *merge* DNA fragment-reads in order to reconstruct the original sequence [35–38]. This task can be broadly categorized into two approaches:

1. *De novo assembly*, for genomes which are not similar to any one sequenced in the past (Fig. 9.17).
2. *Comparative or mapping assembly*, which uses the existing sequence of a closely related organism as a reference during assembly.

The *human genome reference (HGR)* is the keystone to interpret MPS sequencing read data of human DNA samples. The alignment of reads to the HGR is the first step to identify variation of all types including insertions, deletions, number of copies, allele-specific expression, and differential expressions. Misaligning sequences identify structural alterations, while the alignment and analysis of RNA sequence data provide information about gene expression changes.

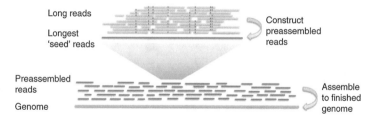

Long reads

Longest
'seed' reads

Construct
preassembled
reads

Preassembled
reads

Genome

Assemble
to finished
genome

Fig. 9.17 De novo hierarchical assembly: Longest reads selected as "seeds," to which all other reads are mapped. A preassembly converts the seed reads into highly accurate preassembled reads to be used for the genome assembly, which are followed by a final consensus-calling step (not shown) (Reprinted from [39], © 2013 with permission from Springer Nature)

However, the human genome reference is still a work in progress [40]:

- The current reference *GRCh38* (released in 2013) is not optimal [41] for some regions of the genome and/or some individuals/ancestries.
- GRCh38 is comprised of DNA from several individual humans.
- Allelic diversity and structural variation present major challenges when assembling a representative diploid genome.
- New technologies, methods, and resources since 2003 have allowed for substantial improvements in the reference genome.
- Additional high-quality reference sequences are needed to represent the full range of genetic diversity in humans.

For the above reasons the NIH funded the *MGI Reference Genomes Improvement Project* aimed at increasing quality and diversity of existing scientific resources. The objective is to sequence and assemble at least five diploid genomes from individuals selected to maximize human genetic diversity (see phylogenetic supertree in Fig. 9.18). All sources have BAC[2] libraries available and whenever possible, samples from a trio (two parents and a child) will be used. Available reference genomes include Puerto Rican, Columbian, Han Chinese, Kinh, Yoruban, Gambian, and three Europeans. Other independent efforts to sequence and assemble new reference genomes include two Japanese, one Malaysian, a Han Chinese, and an Ashkenazim trio.

---

[2] BAC—"Bacterial Artificial Chromosome": Genome is fragmented to the size of bacterial DNA to be vector cloned within a bacterium.

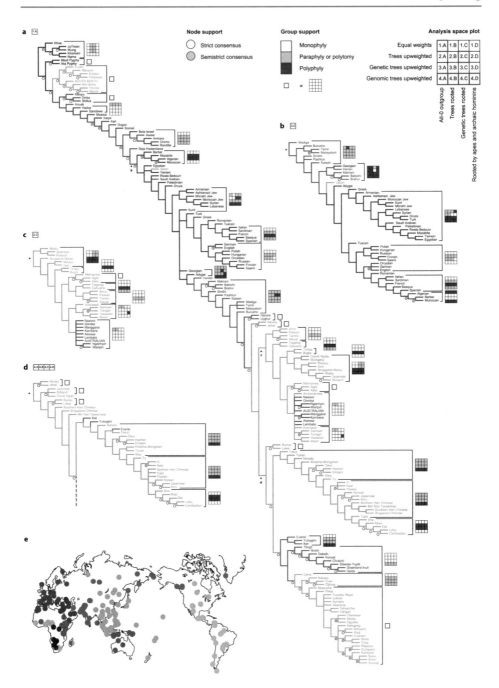

Fig. 9.18 Semistrict consensus phylogenetic supertree of 186 human populations (outgroups not shown) (Reprinted from [42], © 2016 by the Authors, CC-BY 4.0 [3])

9.6 DNA Sequencing by Means of Synthetic Nanopores

Aiming at maximum performances and affordability, solid state pores, that is, fabricated with the standard technological processes of microelectronics, where proposed in 2001 as "envisioned detection devices," and expected to be sensitive, accurate, and suitable for mass production [43].

In this concept, DNA is driven through a synthetic, semiconductor-based nanopore while detecting current fluctuations (Fig. 9.19 left). The first technology proposed for producing such synthetic nanopores is based on *ion beam sculpting* [43]. Depending on the process parameters, a nanopore is either created by etching with an argon ion beam a silicon nitride membrane deposited on a silicon substrate, or an existing, larger pore is narrowed by lateral material transport (Fig 9.20). With these devices, the observation of current blockades, similar to the ones observed by using protein nanopores, was in the focus.

Later, the measurement of transverse tunneling current was proposed [44]. This conceptual nanopore design included embedded emitter and collector tunneling probes, one functionalized with a "phosphate grabber" and the other with a "base reader," and a back gate. The measured tunneling current was expected to specifically differentiate each nucleotide as the DNA is electrophoretically driven through the pore (Fig. 9.19 right).

A big deal of work has been dedicated to the development of a suitable electronic reading, including among others nanowire-based transistors. However, it became clear that the DNA translocation velocity of several mm/s through the pore was much too high; therefore, the attention was placed on the development of a technique to slow down the translocation, similarly to what was done with the biological nanopores, but without using biomolecules. Remarkable results were obtained, among others, first by increasing the

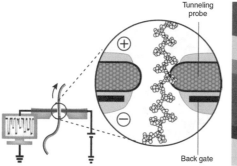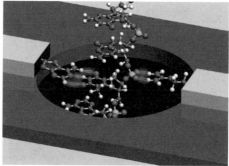

Fig. 9.19 (Left) The "envisioned device": a synthetic solid-state nanopore with embedded readout unit. (Right) Nanopore reader with chemically functionalized probes: As a strand of DNA emerges from a nanopore, a "phosphate grabber" on one functionalized electrode and a "base reader" on the other electrode form hydrogen bonds (purple ovals) to complete a transverse electrical circuit through each nucleotide as it is translocated through the nanopore (Reprinted from [44], © 2008, with permission from Springer Nature)

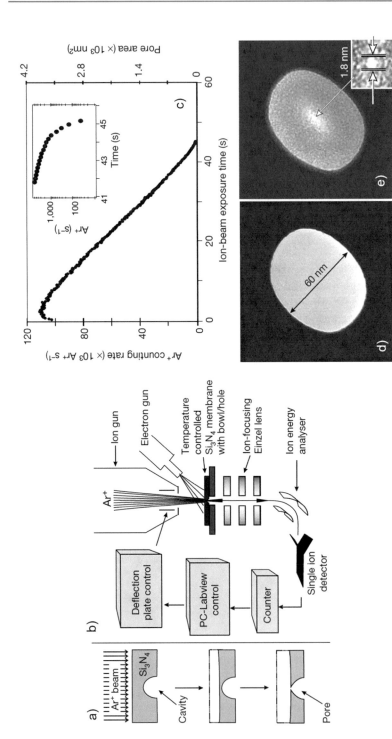

Fig. 9.20 Ion-Beam Sculpting: (**a**) Production of a nanopore by etching using argon ion-beam sputtering. (**b**) Scheme of the procedure, including the control loop. (**c**) Decreasing count rate of transmitted ions is indicative of a narrowing pore by the formation of a nanopore by lateral transport (area given on the right axis). (**d**) TEM image of the initial pore milled by focused ion beam (FIB). (**e**) Final result showing a nanopore of 1.8 nm in diameter (Reprinted from [43], © 2001, with permission from Springer Nature)

electrolyte viscosity [45] and later by depositing an array of nanobeads as frictional structure above the nanopore [46].

However, a reliable direct electronic reading of the DNA sequence has not been achieved so far. Alternatively, a fluorescence-based optical readout was proposed for sequencing DNA as it passes through synthetic nanopores [47]. This method is to date the only one in this category, which reported some evidence of bases identification.

Briefly, in a preliminary preparation step, each nucleotide of the DNA sequence is biochemically converted to two oligonucleotide chains hybridized with complementary oligonucleotides marked with a fluorescent dye and a quencher. Two colors are used in order to identify four binary-coded different bases. When entering the nanopore, the marked DNA oligo is unzipped and the label of the following oligo is unquenched, thus emitting a "photon burst" under laser excitation. Soon after the flashing oligo is unzipped and folds onto itself bringing label and quencher in close vicinity, so that emission stops. This unzipping is at the same time effective in slowing the translocation velocity down to a convenient value for a nearly error-free detection. Two consecutive photon bursts allow identifying the "parent base" in the original strand. Since each nanopore has specific coordinates in the visual field of the EM-CCD, a simultaneous readout of a nanopore array becomes possible (Fig. 9.21).

This approach is properly considered a milestone in this field; however, several tradeoffs like technological difficulties by producing arrays with nearly identical nanopores of 2.2 nm in diameter, the complex binary conversion of the bases, and the corresponding enormous processing time severely limited practical exploitations. Transverse tunneling current remains the most promising detection option; with the help of novel materials and proper surface modifications, synthetic nanopore arrays are expected to achieve the envisioned technological goals.

Question 5

What are the advantages and disadvantages of synthetic nanopores?
How can they be produced?

9.7 Genomics and Personalized Medicine Prospects

The achievements of the Human Genome Project and the wide availability of powerful, affordable NGS tools set strong expectations about their possible impact on the clinical care in the coming future. New terms like *genomic medicine* and *precision medicine* have been coined. So, what can we expect now? Answering this question goes well beyond the scope of this book, however due to its importance we want at least to mention the areas of *Rare Diseases, Prenatal and Reproductive Health, and Cancer*, in which genomic medicine has the potential for providing important results over the entire span of human life [48] (Fig. 9.22).

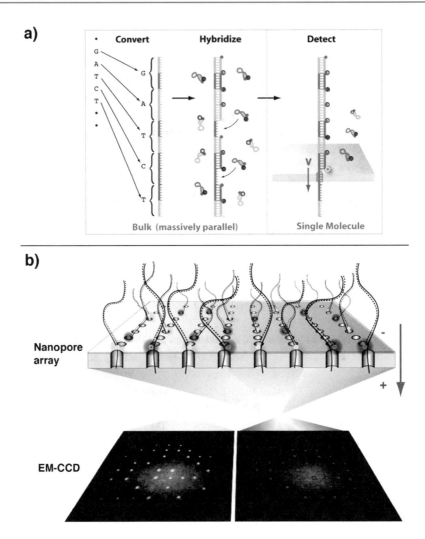

Fig. 9.21 Scheme of the fluorescent readout of DNA sequencing using a synthetic nanopore array. (**a**) DNA conversion and hybridization; photon bursts are emitted when the fluorescent labels are unquenched. This happens when the previous marked oligonucleotide is unzipped. (**b**) Nanopore array imaging: Dichroic filters separate the color information in two distinct fields of the image sensor (Reprinted from [47], © 2010, with permission from the American Chemical Society)

9.8   Supplement: Human Genome Project

The following overview is adapted from the original of the National Human Genome Research Institute (NHGRI [49]).

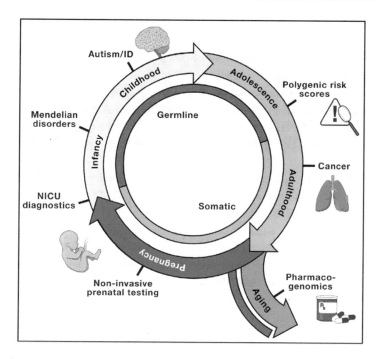

Fig. 9.22 Examples of relevant entry points over the human life span in which genomic medicine has the potential of impacting clinical care (Reprinted from [48], © 2019, with permission from Elsevier)

9.8.1   Yesterday

- A half-century ago, very little was known about the genetic factors in human disease.
- In 1953, J. Watson and F. Crick described the double helix structure of DNA, that is, the genetic instructions for building, running, and maintaining living organisms.
- Methods to sequence nucleotides in DNA were developed in the mid-1970s (Sanger).
- In 1990, the National Institutes of Health (NIH) and the Department of Energy joined with international partners in a quest to sequence all 3 billion base pairs, which is the complete set of DNA in the human body: *The Human Genome Project (HGP)*.
- Its goal was to provide researchers with powerful tools to understand the genetic factors in human disease, paving the way for new strategies for their diagnosis, treatment, and prevention.
- From the start, HGP supported an ethical, legal and social implications research program to address the many complex issues that might arise from this science.
- All data generated by the HGP were made freely and rapidly available on the Internet, serving to accelerate the pace of medical discovery around the globe.
- In April 2003, researchers successfully completed the Human Genome Project, under budget (2.7 billion US$ instead of 3) and more than 2 years ahead of schedule.

9.8.2   Today

- HGP has already fueled the discovery of more than 1800 disease genes, thus today's researchers can find a gene suspected of causing an inherited disease in days rather than the years it took before the genome sequence was in hand.
- There are now more than 2000 genetic tests for human conditions and more than 350 biotechnology-based products resulting from HGP are currently in clinical trials, helping patients to learn their genetic risks and healthcare professionals to diagnose disease.
- A complete genome sequence is similar to having all instructions needed to make the human body. The challenge now is to understand how all of these many, complex parts work together in human health and disease.
- A milestone (2005) was the *HapMap* (http://hapmap.ncbi.nlm.nih.gov/), a catalog of common genetic variation (haplotypes) in the human genome. 2010 HapMap included data from 11 global populations, the largest survey of human genetic variation performed to date. HapMap data accelerated the search for genes involved in common human diseases, yielding already impressive results from age-related blindness to obesity.
- With the drastic drop of sequencing cost, groundbreaking genomic studies are now identifying the causes of rare diseases such as Kabuki and Miller syndromes.
- Pharmacogenomics looks at how genetic variation affects an individual's response to a drug. Pharmacogenomic tests can already identify whether or not a breast cancer patient will respond to the drug Herceptin, whether an AIDS patient should take the drug Abacavir, or what the correct dose of the blood-thinner Warfarin should be.

9.8.3   Tomorrow

- An ambitious new initiative, *The Cancer Genome Atlas* (http://cancergenome.nih.gov/), aims to identify all the genetic abnormalities seen in 50 major types of cancer.
- Understanding disease at the genomic level will yield a new generation of tools, many of which will be drugs, more effective, and with fewer side effects than today's ones.
- NIH-supported access to high-throughput screening of small molecule libraries will provide researchers with powerful new research probes to explore the hundreds of thousands of proteins believed to be encoded by ~25,000 genes in the human genome, provide innovative techniques for development of new, more effective, types of drugs.
- NIH is striving to cut the cost of sequencing an individual's genome to $1000 or less, making it easier to diagnose, manage, and treat many diseases.
- Individualized analysis based on each person's genome will lead to a powerful form of preventive, personalized, and preemptive medicine. By tailoring recommendations to each person's DNA, healthcare professionals will be able to work with individuals to

focus efforts on the specific strategies, from diet to high-tech medical surveillance, that are most likely to maintain health for that particular individual.

- The increasing ability to connect DNA variation with nonmedical conditions, such as intelligence and personality traits, will challenge society, making the role of ethical, legal, and social implications research more important than ever.

Answer 1

DNA sequencing is the determination of the nucleotide sequence in a DNA molecule.

The Sanger method consists of the following steps: (1) Reagents setup. (2) PCR amplification: ddNTPs stop extension randomly, yielding fragments of all possible lengths. (3) Separation of PCR products in a capillary (single lane) gel electrophoresis. (4) Fluorescence readout and data merging into an electropherogram.

Answer 2

During the synthesis, the bases are added one by one and at every step, a certain base-specific signal is detected. This can be a base-specific color of fluorescence, like in the Illumina and Pacific Biosciences methods, or a pH transient, in the ION Torrent system, which is linked to the base solution present in the well in that specific moment in time.

Answer 3

Nucleotide variation	Used in:
1. Deoxynucleotides (dNTPs)	PCR, Sanger
2. Dideoxynucleotides (ddNTPs)	n.n.
3. Dye terminators	Sanger
4. Reversible dye terminators	Illumina
5. 3'-OH unblocked reversible dye-terminators	n.n.
F) Dye-labeled hexaphosphate nucleotides	Pacific Biosciences

The polymerase-mediated synthesis extends the DNA at the 3' position, by binding the OH with a phosphate group. During a normal extension, for example, in the PCR, no cleavage is necessary and the bases bind continuously, one by one, hybridizing the complementary strand.

The phosphoramidite synthesis does not use enzymes; the extension takes place at the 5' position, prior cleavage of the DMTr blocking group. In order to bind, the phosphate of the base to be added is activated with a beta-cyanoethyl group and the coupling reaction requires the presence of triazole. After extension, the beta-cyanoethyl group is removed by oxidation. The resulting DNA at the end of the process is single stranded.

Answer 4

Feature	PacBio	ONT
Generation	Third	Third
Sequencing	By synthesis of complementary strand	By translocation of single strand
Array density	Millions of ZMWs per chip	2 k nanopores per chip
Readout	Fluorescence	Ionic current variations
Read length	Tens of kilobases	Hundreds of kilobases
Reading speed	10 bases per second and ZMW	450 bases per second and nanopore
Portability	No	Yes

Answer 5

Synthetic nanopores have the potential to provide sequencing at very high reading speed and very low cost, thanks to the use of micro- and nanoelectronics.

However, their development is still in progress and a practical exploitation is expected on the mid to long term. The crucial point of the technology consists in the difficulties of controlling the translocation speed of DNA through the nanopores. In addition, the fabrication of arrays with identical nanopores of 2.2 nm in diameter requires further research efforts.

One method of fabrication consists in the use of argon ion beams. By tuning the sputtering parameters, it is possible to work in etching mode for opening pores in a previously thinned membrane, or to work in lateral transport for reducing an initial large pore previously milled by focused ion beam (FIB).

Take-Home Messages
- "Genome Sequencing" means reading base by base the whole genetic DNA present in a cell.
- Sequencing allows to know all genes, that is, not only the genes which are expressed, but also the ones which are not expressed at the time of analysis.
- The Sanger method was the first technique able to sequence DNA strands. It allowed the first complete genome sequencing within the HGP, over a span of 13 years.
- After the success of HGP, new methods able to perform Massive Parallel Sequencing where developed. This is called Next-Generation Sequencing.
- In all NGS systems, the core device is a biochip!

(continued)

- Illumina and ION Torrent are examples of the second-generation sequencing. Both methods require amplification as a prerequisite and perform a sequencing-by-synthesis. Illumina readout is by fluorescence detection, ION Torrent detects pH transients.
- Pacific Biosciences developed the first method of the third generation, able to perform sequencing on a single DNA molecule, that is, without the need of amplification. Its fluorescent detection takes advantage of ZWM arrays.
- Oxford Nanopore Technology sequencing is based on protein nanopores. This technique does not need amplification as well and is able to perform extremely long reads.
- Synthetic nanopores have the potential to boost even more the sequencing speed and lower the analysis costs; however, their development is still far from practical results.
- Assembly is the next step after sequencing, it means to align and merge the data collected from all fragments, check for consistency and finally reach a consensus call.
- Genomics has the potential to impact the clinical care in several genetically linked pathologies, over all ages of human life.

References

1. Sanger F, Nicklen S, Coulson AR. DNA sequencing with chain-terminating inhibitors. Proc Natl Acad Sci U S A. 1977;74(12):5463–7.
2. Estevezj. Sanger-sequencing. https://commons.wikimedia.org/wiki/File:Sanger-sequencing.svg.
3. The Creative Commons copyright licenses. https://creativecommons.org/licenses/.
4. Mardis ER. A decade's perspective on DNA sequencing technology. Nature. 2011;470 (7333):198–203.
5. Head SR, Komori HK, LaMere SA, Whisenant T, Nieuwerburgh FV, Salomon DR, et al. Library construction for next-generation sequencing: overviews and challenges. Biotechniques. 2014;56 (2):61–77.
6. van Dijk EL, Jaszczyszyn Y, Thermes C. Library preparation methods for next-generation sequencing: tone down the bias. Exp Cell Res. 2014;322(1):12–20.
7. Johnson DS, Mortazavi A, Myers RM, Wold B. Genome-wide mapping of in vivo protein-DNA interactions. Science. 2007;316(5830):1497–502.
8. Danker T, Möller C. Early identification of hERG liability in drug discovery programs by automated patch clamp. Front Pharmacol. 2014;5:203.
9. Metzker ML. Sequencing technologies — the next generation. Nat Rev Genet. 2010;11(1):31–46.
10. Rothberg JM, Hinz W, Rearick TM, Schultz J, Mileski W, Davey M, et al. An integrated semiconductor device enabling non-optical genome sequencing. Nature. 2011;475 (7356):348–52.
11. Bustillo J, Fife K, Merriman B, Rothberg J. Development of the Ion Torrent CMOS chip for DNA sequencing. In: 2013 IEEE International Electron Devices Meeting; 2013 9–11 Dec 2013.

12. Mahfooz K, Ellender TJ. Combining whole-cell patch-clamp recordings with single-cell RNA sequencing. In: Dallas M, Bell D, editors. Patch clamp electrophysiology: methods and protocols. New York: Springer; 2021. p. 179–89.

13. Kanagal-Shamanna R. Emulsion PCR: techniques and applications. Methods Mol Biol. 2016;1392:33–42.

14. Zhou Y, Li H, Xiao Z. In vivo patch-clamp studies. In: Dallas M, Bell D, editors. Patch clamp electrophysiology: methods and protocols. New York, NY: Springer; 2021. p. 259–71.

15. Eid J, Fehr A, Gray J, Luong K, Lyle J, Otto G, et al. Real-time DNA sequencing from single polymerase molecules. Science. 2009;323(5910):133.

16. Korlach J, Bjornson KP, Chaudhuri BP, Cicero RL, Flusberg BA, Gray JJ, et al. Real-time DNA sequencing from single polymerase molecules. In: Walter NG, editor. Methods in enzymology. New York: Academic Press; 2010. p. 431–55.

17. Levene MJ, Korlach J, Turner SW, Foquet M, Craighead HG, Webb WW. Zero-mode waveguides for single-molecule analysis at high concentrations. Science. 2003;299(5607):682.

18. Foquet M, Samiee KT, Kong X, Chauduri BP, Lundquist PM, Turner SW, et al. Improved fabrication of zero-mode waveguides for single-molecule detection. J Appl Phys. 2008;103 (3):034301.

19. Korlach J, Bibillo A, Wegener J, Peluso P, Pham TT, Park I, et al. Long, Processive enzymatic DNA synthesis using 100% dye-labeled terminal phosphate-linked nucleotides. Nucleosides Nucleotides Nucleic Acids. 2008;27(9):1072–82.

20. Mitsuhashi S, Matsumoto N. Long-read sequencing for rare human genetic diseases. J Hum Genet. 2020;65(1):11–9.

21. Chen C-Y. DNA polymerases drive DNA sequencing-by-synthesis technologies: both past and present. Front Microbiol. 2014;5:305.

22. Song L, Hobaugh MR, Shustak C, Cheley S, Bayley H, Gouaux JE. Structure of staphylococcal α-hemolysin, a heptameric transmembrane pore. Science. 1996;274(5294):1859–65.

23. Peraro MD, van der Goot FG. Pore-forming toxins: ancient, but never really out of fashion. Nat Rev Microbiol. 2016;14(2):77–92.

24. Meller A, Nivon L, Branton D. Voltage-driven DNA translocations through a nanopore. Phys Rev Lett. 2001;86(15):3435–8.

25. Cherf GM, Lieberman KR, Rashid H, Lam CE, Karplus K, Akeson M. Automated forward and reverse ratcheting of DNA in a nanopore at 5-Å precision. Nat Biotechnol. 2012;30(4):344–8.

26. Manrao EA, Derrington IM, Laszlo AH, Langford KW, Hopper MK, Gillgren N, et al. Reading DNA at single-nucleotide resolution with a mutant MspA nanopore and phi29 DNA polymerase. Nat Biotechnol. 2012;30(4):349–53.

27. Deamer D, Akeson M, Branton D. Three decades of nanopore sequencing. Nat Biotechnol. 2016;34(5):518–24.

28. Timp W, Comer J, Aksimentiev A. DNA base-calling from a nanopore using a Viterbi algorithm. Biophys J. 2012;102(10):37–9.

29. Goyal P, Krasteva PV, Van Gerven N, Gubellini F, Van den Broeck I, Troupiotis-Tsaïlaki A, et al. Structural and mechanistic insights into the bacterial amyloid secretion channel CsgG. Nature. 2014;516(7530):250–3.

30. Brueggemann A, George M, Klau M, Beckler M, Steindl J, Behrends JC, et al. Ion channel drug discovery and research: the automated Nano-patch-clamp technology. Curr Drug Discov Technol. 2004;1(1):91–6.

31. Jain M, Koren S, Miga KH, Quick J, Rand AC, Sasani TA, et al. Nanopore sequencing and assembly of a human genome with ultra-long reads. Nat Biotechnol. 2018;36(4):338–45.

32. He M, Chi X, Ren J. Applications of Oxford nanopore sequencing in Schizosaccharomyces pombe. In: Xiao W, editor. Yeast protocols. New York, NY: Springer; 2021. p. 97–116.

33. Lu H, Giordano F, Ning Z. Oxford nanopore MinION sequencing and genome assembly. Genomics Proteomics Bioinformatics. 2016;14(5):265–79.
34. Garalde DR, Snell EA, Jachimowicz D, Sipos B, Lloyd JH, Bruce M, et al. Highly parallel direct RNA sequencing on an array of nanopores. Nat Methods. 2018;15(3):201–6.
35. Ng PC, Kirkness EF. Whole genome sequencing. Methods Mol Biol. 2010;628:215–26.
36. Sohn JI, Nam JW. The present and future of de novo whole-genome assembly. Brief Bioinform. 2018;19(1):23–40.
37. Armstrong J, Fiddes IT, Diekhans M, Paten B. Whole-genome alignment and comparative annotation. Ann Rev Animal Biosci. 2019;7:41–64.
38. Wee Y, Bhyan SB, Liu Y, Lu J, Li X, Zhao M. The bioinformatics tools for the genome assembly and analysis based on third-generation sequencing. Brief Funct Genomics. 2019;18(1):1–12.
39. Chin C-S, Alexander DH, Marks P, Klammer AA, Drake J, Heiner C, et al. Nonhybrid, finished microbial genome assemblies from long-read SMRT sequencing data. Nat Methods. 2013;10 (6):563–9.
40. Mardis EM. Next-generation sequencing technologies. New York: Springer; 2016.
41. Ballouz S, Dobin A, Gillis JA. Is it time to change the reference genome? Genome Biol. 2019;20 (1):159.
42. Duda P, Jan Z. Human population history revealed by a supertree approach. Sci Rep. 2016;6 (1):29890.
43. Li J, Stein D, McMullan C, Branton D, Aziz MJ, Golovchenko JA. Ion-beam sculpting at nanometre length scales. Nature. 2001;412(6843):166–9.
44. Branton D, Deamer DW, Marziali A, Bayley H, Benner SA, Butler T, et al. The potential and challenges of nanopore sequencing. Nat Biotechnol. 2008;26(10):1146–53.
45. Fologea D, Uplinger J, Thomas B, McNabb DS, Li J. Slowing DNA translocation in a solid-state nanopore. Nano Lett. 2005;5(9):1734–7.
46. Goto Y, Haga T, Yanagi I, Yokoi T, Takeda K. Deceleration of single-stranded DNA passing through a nanopore using a nanometre-sized bead structure. Sci Rep. 2015;5:16640.
47. McNally B, Singer A, Yu Z, Sun Y, Weng Z, Meller A. Optical recognition of converted DNA nucleotides for single-molecule DNA sequencing using nanopore arrays. Nano Lett. 2010;10 (6):2237–44.
48. Shendure J, Findlay GM, Snyder MW. Genomic medicine–Progress, pitfalls, and promise. Cell. 2019;177(1):45–57.
49. The Human Genome Project webpage. https://www.genome.gov/human-genome-project.

Proteome and Membrane Channels

10

Contents

Keywords

2D gel electrophoresis · Protein microarrays · Mass spectroscopy · Membrane channels · Patch clamp

© The Author(s), under exclusive license to Springer Nature Switzerland AG 2021
A. Pasquarelli, *Biosensors and Biochips*, Learning Materials in Biosciences,
https://doi.org/10.1007/978-3-030-76469-2_10

What You Will Learn in This Chapter
In this chapter, you will learn about the methods for analyzing a proteome. After looking at the classical method of 2D electrophoresis, we will explore some concepts and challenges of protein microarrays and discuss different detection methods. Moreover, you will get an introduction to the patch-clamp method for studying the physiology of cells and its use in pharmacology.

    You will be able to explain how protein chips work and their potential use in medical diagnostics. Furthermore, you will be able to describe how to assess efficacy and side effects of drugs by patch-clamping techniques.

10.1 Protein Biosynthesis

The next step in understanding gene function is the examination of the genome at the protein level. In human cells $\sim$400,000 different proteins are produced by the expression of $\sim$25,000 genes. This means that from the same gene a variety of proteins can result. This is controlled by regulatory mechanisms, depending on cell condition and needs.

    Here is a short summary of the protein synthesis steps [1]:

1. *Translation*: Ribosomes assemble polypeptide chains (*primary structure*)
2. *Folding (secondary structure)*
3. *Post-translational modifications*: Involving addition of functional groups, for example:
   - Acetylation: addition of an acetyl group
   - Alkylation: addition of an alkyl group
   - Biotinylation: acylation of conserved lysine residues with a biotin appendage
   - Glutamylation: covalent linkage of glutamic acid residues
   - Glycylation: covalent linkage of 1 to more than 40 glycine residues
   - Glycosylation: addition of a glycosyl group, resulting in a glycoprotein
   - Isoprenylation: addition of an isoprenoid
   - Lipoylation: attachment of a lipoate functionality
   - Phosphopantetheinylation: addition of a $4'$-phosphopantetheinyl moiety
   - Phosphorylation: addition of a phosphate
   - Sulfation: addition of a sulfate group
     - Citrullination or deimination: conversion of arginine to citrulline
     - Deamidation: convert glutamine to glutamic acid or asparagine to aspartic acid
4. *Structural changes (tertiary structure)*
   - Disulfide bridges: covalent linkage of two cysteine amino acids
   - Proteolytic cleavage: cleavage of a protein at a peptide bond
5. *Polymerization* to n-mere proteins (*quaternary structure*)
   - For example: membrane channels, hemoglobin

10.2   Protein Separation with 2D Gel Electrophoresis

The *2D gel electrophoresis,* introduced in 1962 in its early approach [2], reached one decade later the form which is still widely used for proteome analysis [3–5].

This method consists of two steps: The first one separates the proteins by their isoelectric point PI and runs in a strip or tube (Fig. 10.1 left), along a pH gradient superimposed to an electric field (isoelectric focusing or IEF). Here proteins will migrate until they reach a position where pH=PI (net zero charge).

In the second step, the gel of the IEF probe is gently pushed out of the tube and horizontally placed at the top side of a vertically positioned gel slab (Fig. 10.1 right). There, proteins are treated with sodium dodecyl sulfate (SDS) to make them uniformly negatively charged under all pH values and then are separated by mass under the action of an electric field. This second step is also called SDS PolyAcrylamide Gel Electrophoresis (SDS-PAGE). Labeling can be obtained in vivo or in vitro by supplying cells with radioisotope-labeled amino acids before protein synthesis. Alternatively, the sample can be stained in gel after the separation with a fluorescent label like *Coomassie blue*. Labeled proteins are finally detected by autoradiogram or fluorescence respectively (Fig. 10.2).

Pre-separation radiolabeling (Fig. 10.3 left) is therefore particularly suitable for detecting the ongoing protein synthesis on the short term, while post-separation staining (Fig. 10.3 right) gives a measurement of the reference protein amount. These methods can be combined in multiplexed and/or differential experiments where treatments or stimuli lead to changes in protein levels (Fig. 10.4). For a direct comparison, the images are superimposed, prior conversion in false colors. Typical multiplexed experiments (Fig. 10.5) for assessing the complete expression profile consist of eight steps [6].

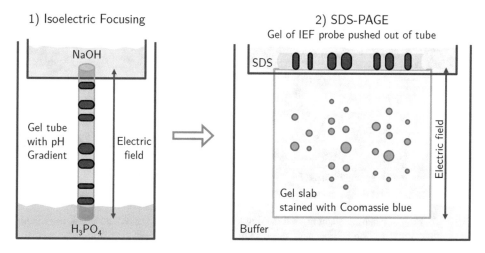

Fig. 10.1 Protein separation with 2D gel electrophoresis: first separation by IEF (left) and second separation by molecular weight by means of SDS-PAGE (right)

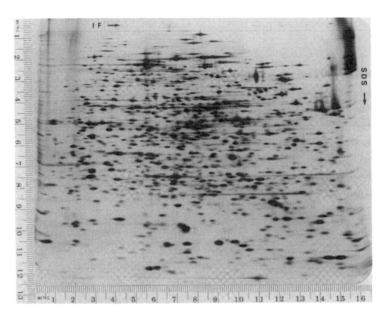

Fig. 10.2 Separation of *E. coli* proteins, labeled with ^{14}C-amino-acids. Approximately 10 μg of protein were loaded onto the gel. The 2D gel was photographed with metric rulers along the two edges of the autoradiogram for establishing a coordinate system, used for tagging the spot positions (Reprinted from [4], © 1975, with permission from Elsevier)

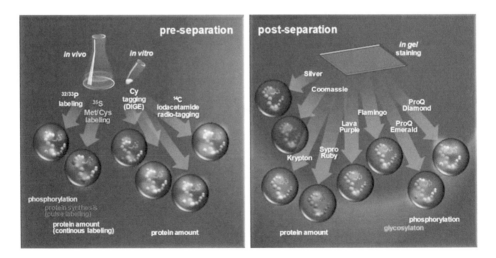

Fig. 10.3 Protein labeling, staining, and tagging techniques for the selective detection of proteins. By multiplexing detection approaches, image analysis may relate different subsets of the proteome and/or different protein levels (Reprinted from [6], © 2007 by the authors, CC-BY 4.0 [7])

Fig. 10.4 Change of protein levels in a heat shock experiment of *Bacillus subtilis*. Green spots (fluorescent stain) represent the reference protein amount, while the red spots (autoradiogram of ^{35}S radiolabel in false color) show the protein synthesis pattern after the thermal stimulus (Reprinted from [6], © 2007 by the authors, CC-BY 4.0 [7])

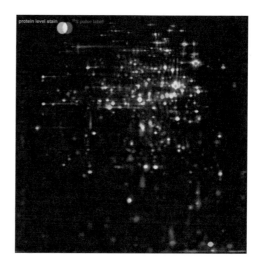

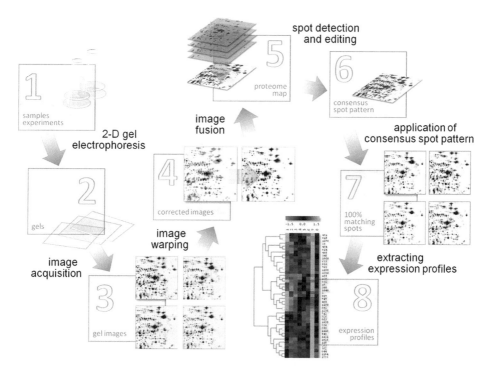

Fig. 10.5 Analysis workflow of a 2D-gel-based experiment: (1) Sample preparation; (2) 2D gel electrophoresis; (3) staining/detection; (4) warping, i.e., alignment of spot positions; (5) images fusion generates a proteome map; (6) construct the consensus spot pattern for the whole experiment; (7) the consensus spot pattern is transferred to all images and subsequently remodeled; (8) expression profiles are extracted and analyzed by clustering, to find relevant proteins (Reprinted from [6], © 2007 by the authors, CC-BY 4.0 [7])

10.3   Protein Microarrays

Protein microarrays (Fig. 10.6 left) are biochips, which measure a variety of proteins [8–10] and work similarly to DNA chips (functionalization, analyte binding, readout). However, there are also remarkable differences:

- Many different kinds of functionalization are possible due to the variety of proteins.
- To date there is no technique equivalent to PCR for amplification of proteins: This leads to enhanced requirements regarding sensitivity, dynamic range, and signal-to-noise-ratio.
- The typical immobilization methods can easily lead to denaturation or inactivation of the capturing proteins (receptors). Hence, new strategies are needed for preserving their native functionality (Fig. 10.6 right).
- Protein 3D structures differ significantly from DNA structures. Fluorophore binding is a probability issue and sophisticated tools for normalization and statistical analysis are required.

The above issues explain the late development of protein microarrays compared to DNA chips.

Question 1
What are the main differences between DNA chips and protein microarrays?

10.3.1 Sandwich Assay Formats

As shown above, analytical protein microarrays are based on the sandwich immunoassay format. Here we describe, compared with *enzyme-linked immunosorbent assay* (ELISA) two methods for the functionalization and readout of such arrays (Fig. 10.7): the *multiplexed sandwich assay* (MSA) and *antibody colocalization microarray* (ACM).

In MSA, different capture antibodies (cAbs) are first arrayed on a slide, then they are incubated with a sample and after rinsing are incubated with a mixture of detection antibodies (dAbs).

In ACM, the initial steps are identical to MSA, but the coordinates of the first spotting are precisely registered and the second spotting delivers each dAb exactly on the coordinates of the spots with the matched cAb, physically colocalizing them. Thus, ACM avoids mix of reagents and looks like an array of singleplex nano-sandwich assays, each requiring only ~1 nL of dAb solution.

MSA is technically simpler than ACM, but is prone to false readings due to a certain binding probability of unspecific dAbs. This is due to the fact that different antibodies may have different specificities and different binding strengths, resulting in a certain degree of "competition" among them. Therefore, for a correct use of this method, it is fundamental to

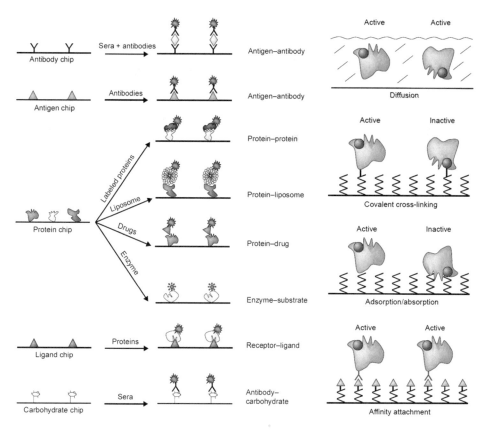

Fig. 10.6 (Left) Variety of protein microarrays: Analytical microarrays involve mostly antibodies or antigens, but also small ligands and carbohydrates are suitable for analyzing specific protein binding activities. Functional protein chips are used for assaying biochemical interactions, e.g., protein–protein, protein–lipid, protein–nucleic acid, enzyme–substrate, and drug target. (Right) Comparison of attachment methods: Proteins are usually randomly oriented, which may alter their native conformation, reduce the activity, or make them inaccessible to analytes. However, by affinity attachment most likely every protein molecule attaches uniformly to the surface remaining in its native conformation, while the analytes easily access the active sites, shown as red dots (Adapted from [8], © 2003, with permission from Elsevier)

assess beforehand the "cross-reactivity maps" among all antibodies and antigens involved in a specific array (Fig. 10.8).

The different specificities and different binding strengths of antibodies lead also to different sensitivities. The consequence is that for a given protein microarray, a calibration plot must be provided for every single antibody employed in that array. The examples given in Fig. 10.9 clearly show how the sensitivity of cAb–dAb pairs to the concentration of the respective specific antigen can differ by several orders of magnitude.

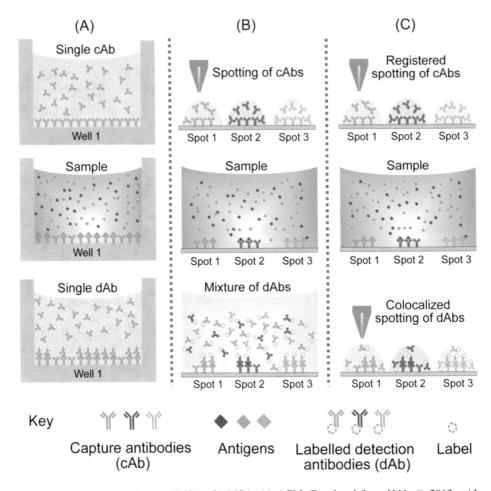

Fig. 10.7 Process flow for (**a**) ELISA, (**b**) MSA, (**c**) ACM (Reprinted from [11], © 2012, with permission from Elsevier)

Once the protein microarray is fully characterized as described above, real measurements, for example, on samples taken from patients, can be done. The data analysis is then performed in a workflow similar to the one already described for DNA chips, consisting of the following:

1. Header with information on experiment setup and imaging software
2. Raw-data analysis, quantification, normalization, quality control, and quality assessment
3. Analysis via database followed by querying and filtering
4. Annotation
5. Clustering

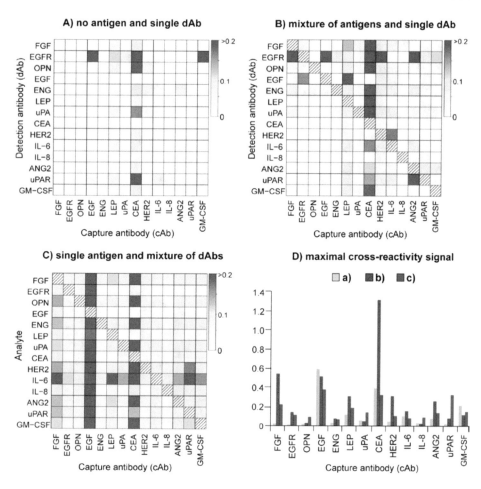

Fig. 10.8 Normalized cross-reactivity maps for an MSA with 14 Ab pairs: (**a**) Fluorescence in a negative control experiment. (**b**) Binding signals obtained by incubation of all arrays with a mixture of analytes followed by a single dAb each. (**c**) Signals observed upon incubating each array with a single antigen followed by a mixture of dAbs. (**d**) Signal comparison for each cAb in (**a**–**c**) (Reprinted from [11], © 2012, with permission from Elsevier)

Data are displayed in a heat map, accompanied by information on the adopted clustering method, for example, a dendrogram in case of hierarchical clustering (Fig. 10.10).

10.3.2 3D-Immobilization

Choosing the right carrier and immobilization method is always a compromise among factors like stability of receptors, function preservation, array density, sensitivity to analyte

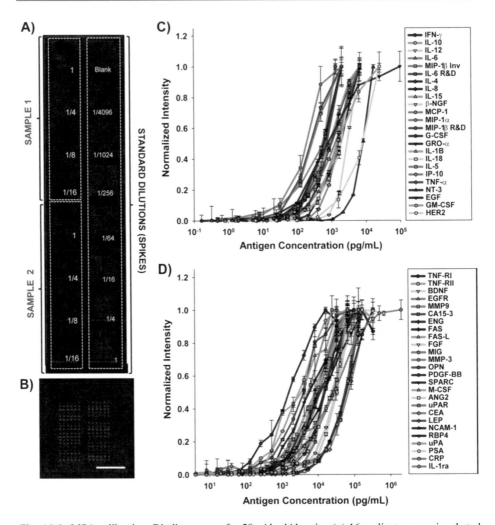

Fig. 10.9 MSA calibration: Binding curves for 50 cAb–dAb pairs. (**a**) 16 replicate arrays incubated with 2 serum samples in 4 dilutions (left), and buffer solution with recombinant proteins in 7 dilutions and 1 negative control used to establish the standard curves (right). (**b**) Detail of a single array with 6 replicate spots for each analyte in a line. Standard binding curves of Ab pairs with (**c**) lower and (**d**) higher sensitivity (Reprinted from [11], © 2012, with permission from Elsevier)

concentration, readout method, and cost. The majority of commercially available microarrays is based on 2D surface chemistry, which may require sophisticated processing techniques for immobilizing the proteins with the correct orientation and without loss of function. Alternatively, a simpler yet reliable immobilization can be provided if the proteins are softly immobilized by photoinduced copolymerization in semispherical polymethacrylamide gel droplets [12]. The pore size allows a good hybridization, since reactants with a molecular mass of up to 150 kDa penetrate readily into gel elements by

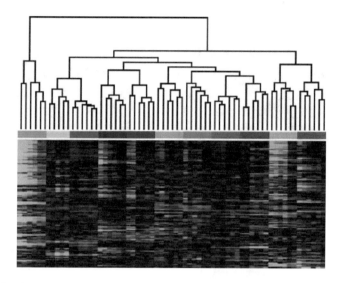

Fig. 10.10 Section of a heat map of a hierarchical clustering of an experiment consisting of 11 individuals with 5 replicate gels each, and 1 average fusion image per individual. Clustering was done for gels (columns) and expression profiles (rows) simultaneously. Gels are color coded by sample, replicates have the same color, sample A is colored in shades of blue, sample B is colored in shades of red. The cluster dendrogram for gels shows that replicates were clustered together, and samples are roughly grouped in the higher-level clusters. The left-most replicate group is probably an outlier, as it branches off early in the dendrogram. The cluster structure in the rows groups proteins with similar expression profiles (row dendrogram not shown). Only about 20% of all expression profiles are shown (Reprinted from [6], © 2007 by the authors, CC-BY 4.0 [7])

passive diffusion. The hydrophilic environment prevents denaturation and loss of function. This immobilization is suitable for label detection as well as mass spectroscopic methods (Fig. 10.11).

Question 2

What are the requisites for a successful immobilization of protein-receptors in an array?

10.4 Discussion on Detection Methods

As for DNA chips, Laser-Induced Fluorescence (LIF) is the preferred method because it is generally safe, simple, very sensitive, has high resolution, and is compatible with standard microarray scanners. However, labeling proteins with fluorophores often yields a reduction of the accuracy, because incorporation of the label may alter the binding properties of the proteins. In some cases, the random labeling of proteins can be addressed with dual-color ratiometric assays, using internal reference spots for each measured target protein.

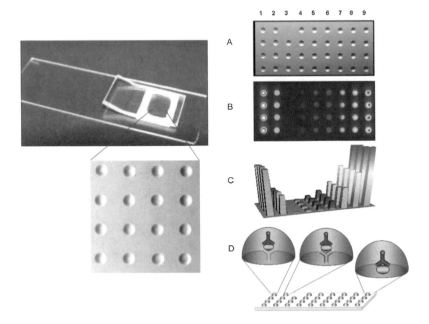

Fig. 10.11 (Left) Biochip for 3D immobilization: (Top) Glass carrier with a transparent reaction chamber; (bottom) magnification showing details of the array with gel drops having 0.1 nL volume and 100 μm diameter. (Right) Quantitative detection of prostate specific antigen split in PSA_{tot} and PSA_{free}. (**a**) Biochip in transmitted light. Columns 1 and 2 contain 200 mg/mL of immobilized monoclonal antibodies (mAbs) to PSA_{tot} and PSA_{free}, respectively; column 3 contains three empty gel elements as control; columns 4–9 contain immobilized PSA (7, 14, 25, 55, 80, and 160 mg/mL PSA, respectively). (**b** and **c**) 2D and 3D computer emulations of the fluorescence image and intensities of spots after the assay. (**d**) Scheme of the reactions on the biochip: (Left and middle) sandwiches formed between capture mAbs, antigen, and detection mAbs. In gel elements used for internal calibration; (right) binary complexes formed between immobilized antigen and detection antibody (Adapted from [12], © 2008, with permission from John Wiley and Sons)

Furthermore, detecting techniques for processing protein microarrays must have a very large dynamic range to avoid saturation at micromolar concentrations, while still being able to detect target proteins at picomolar or femtomolar target levels. Therefore, if detection of copious and scarce proteins is aimed in the same experiment, it is better to work with separate arrays specifically designed for the two ranges. Due to the above issues, there is a significant use of label-free detection methods [13] such as surface plasmon resonance imaging (SPRi), mass spectrometry (MS), and atomic force microscopy (AFM). In the following we give a closer look at MS and AFM.

10.4.1 MALDI-TOF MS

In matrix-assisted laser desorption/ionization time-of-flight mass spectrometry (MALDI-TOF MS) [14, 15] analytes are mixed at molecular level with a thousandfold amount of an organic compound, which acts as energy-absorbing matrix. The wavelength of the pulsed ultraviolet laser is tuned to an absorption level of the matrix; thus, the irradiated energy is transferred to the matrix, which partially vaporizes, carrying intact biomolecules into the vapor phase and positively ionizing them. The released ions are first accelerated in a vacuum tube by an applied voltage (15–25 kV) and then detected, as an electrical signal, at the end of a field-free flight. The time of flight is proportional to the square root of the mass-to-charge ratio of the molecular ions, that is, $t \sim (m/z)^{0.5}$ and is measured with a precision of approximately 10 ppm (Fig. 10.12).

The advantages of MALDI are the spectral simplicity due to single-charged ions, high sensitivity, low noise, low sample consumption, short measurement time, and minimal degradation. Due to the single-charge ionization, the m/z plot directly delivers the mass of detected molecules. In theory, the range of detectable masses is unlimited, however depending on the excitation wavelength, the detector in use and the behavior of the analyte,

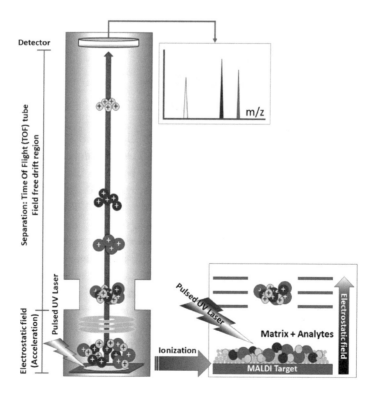

Fig. 10.12 Schematic representation of the MALDI-TOF method (Reprinted from [16], © 2019, with permission from Elsevier)

the maximum measurable mass can be in excess of 1 MDa, which is large enough for dealing with the vast majority of biomolecules and biopolymers. With such large molecular weights, it is more appropriate to use an infrared laser source for excitation, because it significantly decreases the decomposition of the molecules, thus improving the shape of the spectral peak components. However, under such conditions, it is not infrequent to observe multiply charged ions. Under standard conditions, it is generally assumed an optimal mass range between 2 and 20 kDa.

MALDI is a "soft" ionization technique in which the energy from the laser volatilizes the matrix rather than degrade the biomolecules. The preparation of an appropriate polymer–matrix mixture is crucial for good results. The most frequently used matrices are 2,5-dihydroxybenzoic acid (DHB), α-cyano-4-hydroxycinnamic acid (CHCA), sinapinic acid (SA), ferulic acid (FA), and 2,4-hydroxy-phenyl benzoic acid [17].

MALDI applications range not only from biochemistry to microbiology and medicine, but also in organic and industrial chemistry. In the field of biosensing we can mention the following:

- Identification of peptides, proteins, lipids, and oligonucleotides
- Identify pathogens like bacteria, fungi, and viruses
- Study and possibly predict resistance to antibiotics
- Early identify cancer markers

10.4.2 SELDI-TOF MS

Surface-enhanced laser desorption/ionization time-of-flight mass spectrometry (SELDI-TOF MS) can be considered an evolution of MALDI-TOF MS; in fact it is as well a label-free method for the detection of protein biomarkers, based on the m/z ratio, aiming at overcoming the limitation of label-based detection schemes [18, 19]. Actually, SELDI-TOF MS is a proprietary method, integrated in the ProteinChip SELDI system [19–21].

Differently from MALDI, in SELDI the analysis is mediated by the local absorption properties of the chip surface. Indeed, the so-called *bioprocessor* consists of a few arrayed ProteinChips, each one having different surface properties, for example, one is hydrophilic, another is hydrophobic, others provide cation or anion exchange, other spots have surfaces of blank metals, etc., or even a different functionalization like antibodies, enzymes, nucleic acids, etc. In this way, the proteins are characterized not only by mass, but also on the basis of their absorption behavior and affinity binding [22]. The SELDI protocol (Fig. 10.13) includes also the prefractionation of samples. This key feature significantly eases the challenge of analyzing protein-rich fluids like serum, thus allowing a quantitative comparisons of spectral peak intensities between large sets of samples [23, 24].

SELDI systems are tailored for quickly analyzing and comparing multiple biological samples, with a particular capability of detecting altered protein expression profiles in the

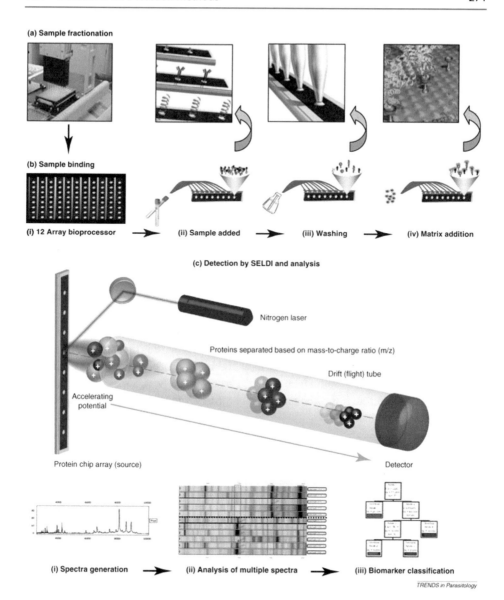

Fig. 10.13 Overview of the SELDI protocol for discovery of diagnostic biomarker of Chagas disease [24], showing that SELDI has the potential to predict progressive disease leading to cardiopathy. (**a**) Sample fractionation performed using anion exchange resin and a pH gradient. (**b**) Sample binding: (i) ProteinChip array with 96 spots; (ii) sample binds to the ProteinChip array surface; (iii) unbound proteins are washed off the array; (iv) matrix (e.g., sinapinic acid) applied to each spot. (**c**) Detection by SELDI and analysis: (i) Spectra generation: pooled and replicate samples are also used for quality control; (ii) Spectra of serum samples from 11 patients. The top 5 samples are from chronic Chagas patients with electrocardiogram (ECG) alterations, the bottom 6 are from patients with normal ECG; (iii) biomarker classification and generation of decision tree based on diagnostically relevant biomarkers (Reprinted from [27], © 2010, with permission from Elsevier)

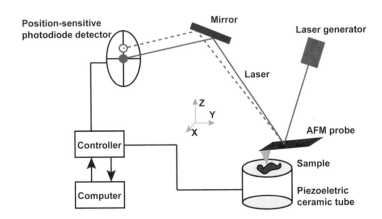

Fig. 10.14 Schematic diagram of AFM working principles (Reprinted from [29], © 2018 by the authors, CC-BY 4.0 [7])

various phases of a disease, that is, development, progression, and therapy [25, 26]. This approach is called *protein profiling*.

Serum samples are spotted onto such bioprocessor and unbound material is washed off. Then the matrix solution is dispensed on the same spots. After this preparation, the chip is irradiated with a laser, resulting in soft release and ionization of the captured molecules. The ions travel through a vacuum tube and their mass-to-charge ratios are calculated from their time of flight through the vacuum chamber. SELDI is best suited for biomolecules with a mass smaller than 20 kDa.

10.4.3 Atomic Force Microscopy (AFM)

Atomic Force Microscopy (AFM) uses the topological changes of the surface to identify the captured proteins (e.g., on an antibody array), which means AFM detects the increase in height, and thus it is able to measure quantitatively the binding interactions (Fig. 10.14). Moreover, the sub-nanometric resolution of AFM allows imaging and resolving complex conformations of proteins and their aggregation at the molecular scale having the potential to unveil key aspects of pathologies like neurodegenerative disorders and cancer [28, 29].

Due to the high spatial resolution of AFM, with this method the spots require a size of just a couple of micrometers, and therefore it is possible to array a large number of spots on a single slide. The drawback is that since the AFM linear scanning is relatively slow, the total readout time can be very long and is proportional to the area of the array. Figure 10.15 illustrates the measurements of rabbit antibodies against goat antibodies immobilized in spots on a gold-coated silicon substrate.

The AFM technique can be adopted in a variety of mechanical biosensing approaches [31]. Here we briefly mention a few of them:

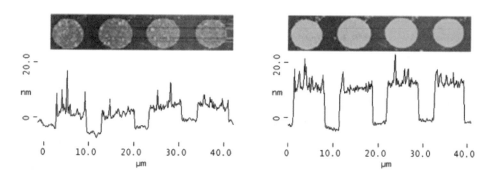

Fig. 10.15 AFM topographic images and height profiles of protein–protein affinity binding: (Left) Spots with immobilized goat IgG. (Right) Same spots after incubation with rabbit anti-goat IgG. After binding, the height of the spots increases from about 6 nm to about 12 nm, which is consistent with the size of antibodies (Reprinted from [30], © 2004, with permission from John Wiley and Sons)

- Functionalization: The tip can carry and deposit receptor molecules thus realizing an array of nanospots. This method is called *Deep Pen Nanolithography* (DPN).
- Ligand-functionalized tips can measure the binding forces between receptors and ligands by approaching the receptor thus allowing bond formation and then retracting the tip until rupturing the bond.
- Measure stiffness of cells membrane and changes thereof due to ageing or pathologies. This can be performed with tips of different shapes (e.g., pyramidal, conical, cylindrical, spherical etc.).

Question 3
Explain the MALDI-TOF MS method. What are its key features?

10.5   Commercial Microarray Technologies

Several commercial protein microarray have become available in the last two decades; listing all of them goes beyond the scope of this book, therefore interested readers are redirected to [10, 32]. However, it is worth mentioning a few concepts, which may provide some guidance in this exploration.

Classification   Protein microarrays are divided in three main classes:

- Functional: to probe the biochemical properties of the proteins in the sample.
- Analytical: to recognize and quantify proteins in the sample, by affinity binding.
- Reverse phase: complex samples are immobilized on a carrier, then known affinity reagents are incubated and the presence of target molecules is detected.

Model Organism A given set of proteins for microarray functionalization is specific to the proteome of a certain organism to be studied, for example, a virus or bacterium or even a mammalian.

Expression System The proteins for microarray functionalization are not always taken from their natural source. Especially proteome microarrays of complex organisms are often assembled by genome-wide cloning into expression vectors, that is, the proteins are synthetized in bacteria like *E. coli* or eukaryotic cells like *S. cerevisiae* (budding yeast) after transfection. However, in such systems the proteome coverage is not complete. Moreover, if the post-translational modifications are missing in the chosen expression system, the variety of expressed proteins is low. For the human proteome, *S. cerevisiae* demonstrated to be one of the best heterologous expressions systems. It is commercially available under the trademark HuProt [33] and has a coverage in excess of 21,000 proteins.

10.6 Membrane Channels

Membrane channels enable the flow of ions and molecules through the cell wall (membrane). Ions and molecules passively diffuse along the concentration gradient, so it is not an active transport and therefore it does not consume energy.

Membrane channels can be classified in two groups:

- *Ion channels* (e.g., Na^+, K^+, Ca^{2+}, Cl^-) are ion specific, which means only a specific type of ion can pass through a specific membrane protein (channel). They are generally gated (opened/closed), depending on the channel type, by different signals (voltage, light, chemical substances, pressure). Because of the relevance of membrane proteins for cell physiology, ion channels are investigated in devices implementing the *patch-clamp* technique.
- *Pores*: These membrane proteins are found in bacteria. Pores are always open (non-gated) and allow ions and small molecules such as water, sugars, and amino acids to flow along the concentration gradient. Because pores allow several similar chemical species to pass through, they are less selective. But there are also very selective pores for just one particular molecule. For the osmotic regulation there is an adequate pore expression (aquaporin, OmpF) from the cell to regulate the water flow. Other pores are used for sugar transport (maltoporin) or as toxins especially by pathogenic bacteria (e.g., α-hemolysin from the bacterium *Staphylococcus aureus*, the protective antigen of the anthrax pathogen *Bacillus anthracis*, pneumolysin of the bacterium *Streptococcus pneumoniae*, etc.). This family of membrane channels is used in nanopore devices (see Chap. 9).

10.6.1 Ion Channels

The physiology of electroactive cells (e.g., neurons and muscle cells) is largely based on controlling/changing the potential difference between the interior of the cell and the external environment, this is called *transmembrane potential*. Both potential values result from the local sum of the contributions of each ion type and can be calculated, for example using the Nernst's equation.

For each individual type of ion, the specific regulation of ion exchange occurs at the cell membrane while electrophysiological cell activity is ongoing. Here, Na^+, K^+, Ca^{2+}, Cl^- are the most important ions. These ions pass the membrane through specific ion channels, which may be open (activated) or closed (deactivated). Therefore, these channels are called *gated channels*. Their opening and closing can occur in response to voltage change (voltage-gated channels) or receptor activation by specific ions or chemical ligands (receptor-gated channels). Channels activated by light (photosensitive) or controlled by mechanical stress (i.e., pressure) also exist, but are less common (Fig. 10.16).

Ion channels are involved in several physiological processes. Here are just a few examples:

- Communication among cells and with the environment:
  - Action potentials
  - Synaptic transmission
- Muscle contraction:
  - Heart activity
  - Bowel activity

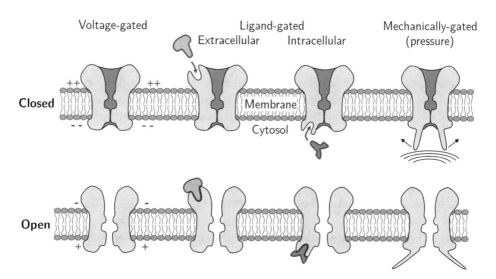

Fig. 10.16 Cartoon illustrating the fundamental gating mechanisms of ion channels

– Skeletal muscle activity
• Biochemical and environmental variables:
  – pH regulation
  – Taste and odor sensing

10.7   Patch-Clamp Technique

The Patch-Clamp technique [34, 35] is the gold standard for studying the behavior of voltage-gated ion channels. The principle of operation consists in attaching a cell with a micropipette, having an opening at the tip of approximately 1 μm in diameter, then apply an appropriate underpressure for obtaining a tight contact with the cell membrane. At this point, the experimenter has the option of working under this *cell-attached* configuration, aiming at investigating a single channel, or apply a bit stronger underpressure for rupturing the cell membrane, thus gaining access to the intracellular environment. The latter is the *whole-cell* configuration for investigating the cumulative behavior of all channels (Fig. 10.17).

There are also other configurations like *inside-out* and *outside-out* for studying single channels on a small excised patch of the cell membrane, either from the inside or from the outside. However, those configurations are less frequently adopted and require highly skilled operators. Before starting a measurement protocol, it is important to check the impedance of the micropipette and in particular to assess the value of the leakage component R_{leak}, whose optimal value ranges around 1 GΩ or above. If this value is significantly smaller, the patch is leaky, resulting in large artefacts. The figure below shows the cell-pipette system and the equivalent impedance model (Fig. 10.18).

Once the desired configuration is achieved, two basic option are available for the measurements:

Voltage Clamp The transmembrane potential is imposed (clamped) by the instrument (V_{clamp}) and the resulting current transients are measured (Fig. 10.19 b). This method is used to assess the gating thresholds of the channels and the time course of the resulting currents under different conditions. Practically, a series of rectangular voltage pulses of increasing intensity is delivered to the cell. Every pulse starts usually from a hyperpolarization potential, for example, −70 or −80 mV, for keeping all channels closed, and rises by a few tens of mV. Such set of pulses is usually represented overlaid in a ladder fashion for compactness. Also, the recorded current signals are displayed overlaid in a single plot.

Current Clamp Conversely, in these measurements current transients are imposed by the instrument and resulting variations of the transmembrane potentials are detected (Fig. 10.19 a). Aim of this approach is eliciting action potentials from electroactive cells while studying the influence of external variables like temperature, drugs, neurotransmitters, toxins, etc. on

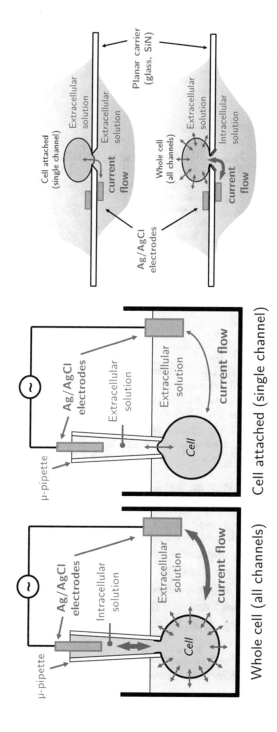

Fig. 10.17 Schematic illustration of the two most common operation schemes of the patch-clamp method, namely *whole cell* to investigate the cumulative behavior of all channels, and *cell attached* to focus on one single channel. The solutions at the sides of the investigated channels must provide the physiological gradient for proper operation. (Left) Classical approach with a glass micropipette. (Right) Technological implementation on a planar carrier after milling a micro-hole. In most patch-clamp configurations only two electrodes are used, therefore they must be both of type II (i.e., non-polarizable)

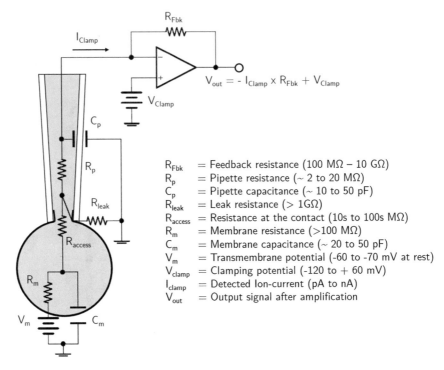

$$V_{out} = - I_{Clamp} \times R_{Fbk} + V_{Clamp}$$

R_{Fbk} = Feedback resistance (100 MΩ – 10 GΩ)
R_p = Pipette resistance (~ 2 to 20 MΩ)
C_p = Pipette capacitance (~ 10 to 50 pF)
R_{leak} = Leak resistance (> 1GΩ)
R_{access} = Resistance at the contact (10s to 100s MΩ)
R_m = Membrane resistance (>100 MΩ)
C_m = Membrane capacitance (~ 20 to 50 pF)
V_m = Transmembrane potential (-60 to -70 mV at rest)
V_{clamp} = Clamping potential (-120 to + 60 mV)
I_{clamp} = Detected Ion-current (pA to nA)
V_{out} = Output signal after amplification

Fig. 10.18 Equivalent circuit of the cell-pipette system in whole-cell configuration connected in voltage-clamp mode with a transimpedance amplifier. The typical values of the listed parameters depend on the cell type and the adopted patch-clamp configuration (i.e., whole cell, cell attached, etc.). The leak resistance R_{leak} beside the pipette is due to the quality of the seal. This should be larger than 1 GOhm for good measurements. For optimal signal fidelity and artifact minimization, the capacities are compensated with an appropriate circuit available in the front end (not shown)

this activity. The results can shed lights on the influence of external variables on healthy cells or help in finding a therapy in case of channels malfunction due to pathologies.

In both patch-clamp methods, depending on the type of cells and ion channels being investigated, the signals are usually recorded in a bandwidth from DC to 10–25 kHz. The sampling frequency is typically four times the bandwidth. For preserving the signal morphology, it is important to use low-pass filters having a constant group delay, that is, linear phase shift. Bessel filters of the fourth to eight order are an appropriate choice at the analog signal level. Further filtering in the digital domain can be included in the data-processing algorithms.

Ion channels are important targets for drug screening, because their function/dysfunction have a strong impact on cellular activity and signaling. Two goals are particularly important:

- Discover new drugs to address genetic diseases involving ion channels mutations.
- Test new drugs for side effects on electrophysiological activity of ion channels.

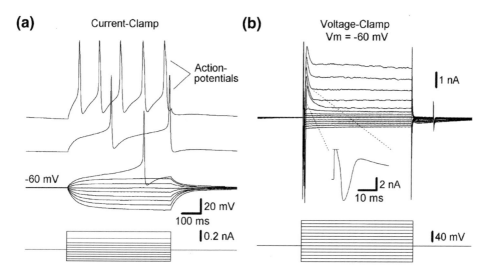

Fig. 10.19 Whole-cell recordings from a neuron on a planar patch-clamp chip. (a) Current-clamp measurements: Voltage responses (top) to a graded series of intracellular current pulses (bottom) showing the threshold for eliciting action potentials. The two upper curves are shifted up for clarity. The current pulses were applied at a membrane potential of -60 mV ($I_{clamp} = -45$ pA increase/step). (b) Current responses (top) of the same neuron to a series of intracellular voltage pulses (10 mV increase/pulse) from a membrane potential of -60 mV. The inset shows a region with an expanded time scale (Reprinted from [36], © 2010, with permission from Springer Nature)

Table 10.1 Examples of pathology-linked ion channels

Channel	Significance (pathology)
hERG	Long QT syndrome
Kv1.3	Target for immunosuppressants
Nav1.5	Long QT syndrome
AChR_7	Inflammation
CLC7	Osteoporosis
P2X7	Immune disease
TRPC6	Asthma
TRPM8	Prostate cancer

Here below are a few examples of ion channels whose malfunction in associated to pathologies. In the following, we will focus on the relevance of the hERG [1] channel in the process of drugs screening (Table 10.1).

---

[1] Human Ether-a-go-go Related Gene (hERG) encodes the Kv11.1 potassium channel, which is responsible for the repolarizing current in the cardiac action potential. The vast majority of drugs associated with acquired long QT syndrome are known to interact with the hERG potassium channel.

Question 4

What variables can be assessed with the patch-clamp method?

10.8   Ion Channels and Cardiac Physiology

Cardiomyocytes (heart muscle cells) possess calcium, potassium, and sodium ion channels, which work synchronously in specific time segments of a heartbeat [37]. Furthermore, for every ion species there are few different types of channels, with different behaviors and sensitivities. More specifically, different types of K-channels play an important role for maintaining the resting membrane potential (RP) and by repolarizing the cell at the end of the action potential (AP). Opening and closing of ion channels alter specific ion conductances with the consequent ion flows, thus determining the membrane potential at every moment in time of the cardiac cycle.

The transmembrane potential V_m relative to the instant ion concentrations can be calculated by means of the Goldman–Hodgkin–Katz voltage equation [38, 39], taking into account the contributions of the three above-mentioned ion species, plus chlorine which is present in both the intracellular and extracellular environments (Fig. 10.20):

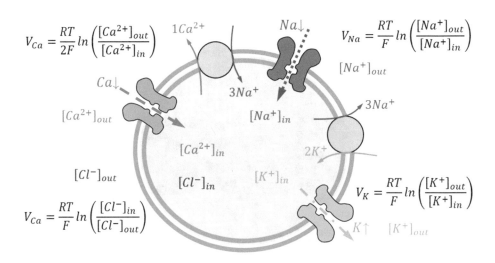

$$V_{Ca} = \frac{RT}{2F} \ln\left(\frac{[Ca^{2+}]_{out}}{[Ca^{2+}]_{in}}\right)$$

$$V_{Na} = \frac{RT}{F} \ln\left(\frac{[Na^+]_{out}}{[Na^+]_{in}}\right)$$

$$V_K = \frac{RT}{F} \ln\left(\frac{[K^+]_{out}}{[K^+]_{in}}\right)$$

$$V_{Ca} = \frac{RT}{F} \ln\left(\frac{[Cl^-]_{in}}{[Cl^-]_{out}}\right)$$

Fig. 10.20 Simplified representation of the individual contributions of the various ionic species to the membrane potential. Here the round structures represent the ion pumps. For every formula it is assumed a steady state, that is, no ionic current flowing and a permeability of 1 for the given ion species and 0 for the others. However, in the real case the ions flow, thus leading to a deviation from this simplified scheme. The respective ion concentrations and the resulting potential values change constantly, with cyclic progression, during the heartbeats

Fig. 10.21 Ion flow and phases of the ventricular action potential in a cardiomyocyte

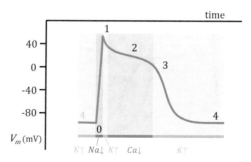

$$V_m = V_{in} - V_{out} = \frac{RT}{F} \ln \left(\frac{P_K[K^+]_{out} + P_{Na}[Na^+]_{out} + P_{Ca}[Ca^{2+}]_{out} + P_{Cl}[Cl^-]_{in}}{P_K[K^+]_{in} + P_{Na}[Na^+]_{in} + P_{Ca}[Ca^{2+}]_{in} + P_{Cl}[Cl^-]_{out}} \right)$$

where P_X represents the membrane permeability to the X-ions and the values in brackets the respective ion concentrations at the sides of the membrane. Please note that the calcium contributes with double charge and chlorine, being negatively charged, contributes in opposite way to V_m.

The above equation is valid under equilibrium conditions, that is, without ion flows. In real cases the ions flow intermittently in one direction through the ion channels and continuously in the opposite directions through the ion pumps, which work for restoring the ion concentrations. These activities lead to some deviation from the simplified formula above. For example, calculating the resting potential with the physiologic levels of $[K^+]_{in} = 150$ mM and $[K^+]_{out} = 4$ mM, results in -96 mV under equilibrium conditions, but in the real case, the measured value is -90 mV. Opening and closing sequence repeats cyclically at every action potential (Fig. 10.21).

As shown in Fig. 10.21, when an action potential is elicited in a cardiomyocyte (0), K^+ channels close and the quick Na^+ channels open shortly. This results in an abrupt depolarization. Immediately after (1), within a few milliseconds Na^+ channels close and rapid K^+ channels produce a short initial repolarization, then Ca^{2+} channels open (2) to maintain a depolarized state. It follows an inactivation of Ca^{2+} channels and a reopening of K^+ channels (3), leading to membrane repolarization (4). In the background, ion pumps work continuously for restoring the respective ion concentrations.

Most antiarrhythmic medications used in the therapy of cardiac arrhythmias target malfunctioning ion-handling systems [40].

10.8.1 The Long QT Syndrome

At a macroscopic level, the Long QT syndrome (LQTS) is observable in the electrocardiogram (ECG). Here the wave components of a heartbeat are referred to by respective letter

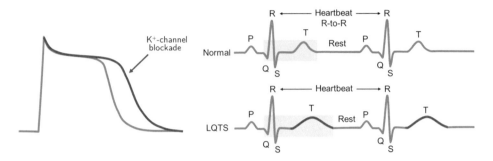

Fig. 10.22 (Left) Ventricular Action Potential: A delayed repolarization due to the blockade of the hERG potassium channel leads to LQTS. (Right) LQTS evidence in the ECG: (Top) Under normal QT conditions the TP resting interval is long enough for supplying nutrients to the heart. (Bottom) The shortened resting time observed in a patient with LQTS, leads to a permanent stress of the cardiac cells, which can lead in the worst cases to ventricular fibrillation and sudden death

names: P, Q, R, S, and T. LQTS appears as a prolonged interval from the beginning of the Q-wave to the end of the T-wave, whereas the overall duration of the heartbeat remains in the normal range (Fig. 10.22). The LQTS can be either congenital due to genetic mutation or acquired after drug intake; the latter is usually reversible.

At the microscopic level, the blockade of the hERG channels (K^+) is a common cause of the LQTS. Such blockade leads to a delayed repolarization of the cardiomyocytes. The consequence is that the resting time between the end of the T-wave and the onset of the next P-wave is shortened. As a result, the heart is in a condition with an increased risk of abnormal heart rhythms like ventricular tachycardia, especially under physical exercise or mental stress. If such condition becomes severe, the heart starts beating erratically (ventricular fibrillation) with a negligible net blood transport. This state can lead to fainting (syncope), drowning (suffocation), or even sudden death. For an in-depth study of the hERG channel the interested readers are redirected to [41].

Question 5

Which kind of membrane channels are involved in cellular electrophysiology?
    What are the most important channels in cardiomyocytes?

10.8.2 hERG Pharmacology

The blockade of the hERG channels is often caused by the intake of drugs or drug combinations. The antihistamine drug terfenadine is an example of such a drug. In the 1990s, it was replaced by fexofenadine. However, other medications such as flunarizine (for migraine), cisapride (for reflux esophagitis, also called heartburn), ketoconazole (antifungal), propafenone (antiarrhythmic), or cerivastatin (cholesterol reducing) in combination with gemfibrozil (generic lipid-lowering) can cause a blockade of the hERG

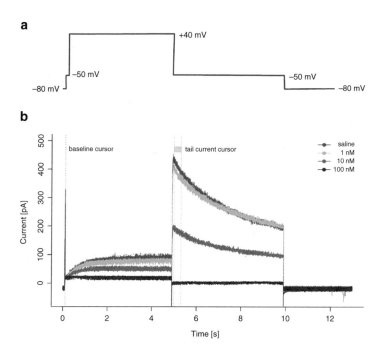

Fig. 10.23 hERG channel blockade caused by Cisapride: (**a**) Stimulation protocol for measurements of hERG channel activity. (**b**) hERG patch-clamp recordings showing the effect of Cisapride. Four representative traces from the same cell are superimposed to visualize the effect of increasing concentrations of the known hERG blocker Cisapride (Reprinted from [44], © 2014 by the authors, CC-BY 3.0 [7]).

channels. A comprehensive and maintained list of compounds that show reduction of the hERG channel activity is available at [42]. Due to the importance of this topic, in 2020 a systematic strategy was defined with the consensus of the most important companies and research institutions active in the field [43]. Since the early 2000s, all new medications need to be tested before approval for possible side effects on the hERG channel. This is done by patch-clamp as shown in Fig. 10.23 for the drug Cisapride.

10.9 Instrumentation

There are two basic concepts of patch-clamp instruments: *manual* and *automated* operation.

Manual operation is the traditional method, it requires skilled operators and allows to run high-quality measurements rich of fine details with the possibility of flexible protocols, for example, by combining patch-clamp with fluorescence and even single-cell mRNA sequencing [45] or also by performing in vivo studies [46]. However, manual operation is quite slow, having a throughput of up to about ten experiments per day. Beside the classical

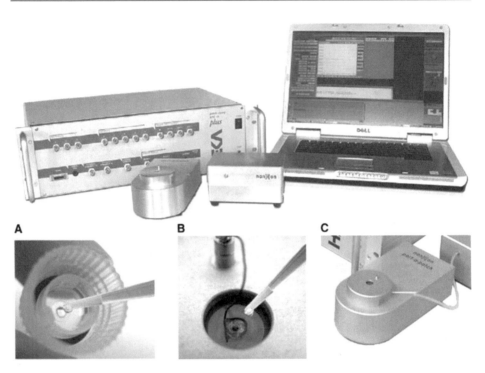

Fig. 10.24 (Top) The first generation of the single-cell patch-clamp system "port-a-patch" of Nanion Technologies with an amplifier of Heka Elektronik. (Bottom) Close-up view of the chip from the inside (**a**), from the top side after assembly (**b**), and the tailored mini Faraday cage (**c**) (Reprinted from [51], © 2008, with permission from Springer Nature)

method based on glass micropipettes, there are also planar-chip approaches for manual operation, which significantly simplify the work, thus making patch-clamp suitable for practical courses and basic research (e.g., at universities). An example of such systems is the *port-a-patch* [47] from Nanion Technologies (Fig. 10.24), which works in tandem with standard amplifiers for patch-clamp and electrophysiology (e.g., from companies like Heka Elektronik [48], Molecular Devices [49], and NPI Electronic [50]).

Automated operation is the other approach. This is particularly suitable for application scenarios necessitating a high throughput rate, for example, in case of drug discovery/drugs screening. Such systems work fully automated with high parallelism, thus allowing to examine the effect of several thousand molecules (drug candidates) per day on the physiology/pathology of the cells. In this class of instruments, there was a great deal of development in the early 2000s, especially from companies like Cytocentrics, Fluxion Biosciences, Flyion, Molecular Devices, Nanion Technologies, and Sophion Bioscience; however, most products did not have commercial success and were discontinued [52].

At present there are only two major players in this field, namely Nanion Technologies [53] and Sophion Bioscience [54]. Both companies offer a full range of automated devices

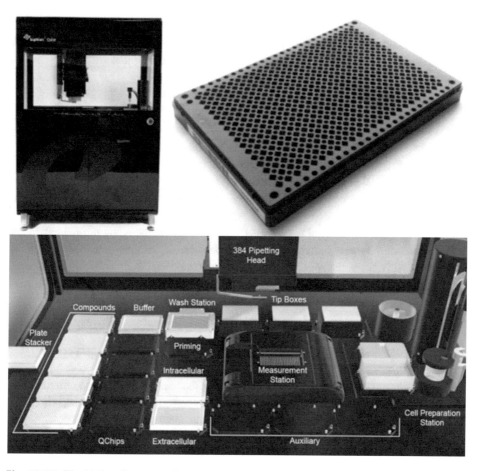

Fig. 10.25 The high-end automated system Qube 384 from Sophion Biosciences: (Top) the whole apparatus (left) and the QChip 384 measurement plate (right). (Bottom) Detailed view of the workplane. Compound plates are placed in the left row, QChip 384 measurement plates are placed in the second row from the left and the intracellular and extracellular solutions are placed in the third row from the left. The 384 robot pipette tips are placed in tip boxes behind the measurement station. The cell preparation station is situated on the right side of the workplane (Reprinted from [55], © 2021, with permission from Springer Nature)

ranging from 16 to 384 channels, with the possibility of twinning two machines for a total of 768 channels, all of them based on planar chip arrays having the same form factor of microtiter plates (Fig. 10.25).

Question 6

What is the role of patch-clamp in the workflow of drugs discovery?

Answer 1

Protein microarrays share many properties of DNA chips like the working concept in three steps (functionalization, incubation, readout); however, there are also remarkable differences, as follows:

- DNA is always a linear chain, while proteins have complex 3D structures, which make the immobilization and the readout of fluorescence emission more challenging tasks.
- Preserving the functionalization can be challenging as well.
- For proteins, there is to date no amplification method available. Therefore, working with low analyte concentration requires a detection method with a high signal-to-noise ratio.
- Protein microarrays require a complete characterization including normalization and calibration for every individual protein to be detected, in order to assess the individual sensitivity for the specific analytes and the cross-reactivity with other molecules in same array.

Answer 2

- Avoid denaturation and preserve functionality, that is, binding specificity and strength.
- Provide convenient orientation, thus allowing the analyte to access the binding site.
- Fluorophore binding should not interfere with the active binding site and should be positioned conveniently for a good efficiency of the fluorescence readout.
- Receptors should remain fully functional and reliably bound to the carrier plate even after several measurement/regeneration cycles.

Answer 3

This is a mass spectrometry method. Proteins are mixed with a thousandfold amount of matrix. A laser beam, with a suitable wavelength for good absorption, is pulsed on the matrix, which vaporizes while transporting and ionizing analyte molecules. Such charged molecules are accelerated in a linear electric field and reach a detector after a certain time-of-flight (TOF), which is measured. The TOF is proportional to the mass–charge ratio. In MALDI, the molecules are single-charged, which makes the mass calculation straightforward.

Key features of MALDI are follows:

- Optimal detection range is 2 and 20 kDa, but can detect up to 1 MDa and beyond.
- Minimal degradation of analyte molecules.

– The method is fast and can detect a variety of analytes: peptides, proteins, lipids, oligonucleotides, bacteria, fungi, and viruses, and predict resistance to antibiotics and identify cancer markers.

Answer 4

The primary variables are current and potential transients, depending on the chosen configuration, that is, voltage clamp and current clamp, respectively. Also, the time duration of the transients and the thresholds for eliciting activities are important assessments. Moreover, when delivering compounds like activators or inhibitors, it is possible to detect the sensitivity of specific channels to those compounds.

Answer 5

Voltage-gated ion channels are directly driven by voltage signals.

In the cardiomyocytes, three groups of channels work synchronously and cyclically together during the action potential. These are, in the timing sequence, sodium, calcium, and potassium channels.

Answer 6

Patch-clamp is one of the fundamental methods in the drug-screening processes. There are two main goals of such investigations:

– Discover new drugs to address genetic diseases involving ion channels mutations.
– Test new drugs for side effects on electrophysiological activity of ion channels.

Take-Home Messages
• Analyzing the gene expression at the protein level has the potential to deliver more details than the DNA-analysis alone.
• Traditionally, protein synthesis is analyzed with the 2D electrophoresis; multiplexed and differential protocols allow following the effect of external variables on the cells.
• Protein microarrays are the modern biochip alternative to the above method. The approach is similar to the one of DNA chips, but there are also remarkable differences due to the variety and complexity of proteins.
• Fluorescence is the most commonly adopted readout method; however, in some cases mass spectrometry and AFM detection can provide superior performances.

(continued)

- Membrane proteins like ion channels are investigated by means of the patch-clamp methods.
- Planar patch-clamp biochips have been successfully developed and arrayed in complexity of up to 384 channels per plate. Such sensors are utilized in fully automated high throughput systems.
- Patch-clamp is one of the fundamental methods in the study of electrophysiology and in the drug discovery/drug screening processes.

References

1. Pasquarelli A. Biochips: technologies and applications. Mater Sci Eng C. 2008;28(4):495–508.
2. Raymond S, Aurell B. Two-dimensional gel electrophoresis. Science. 1962;138(3537):152.
3. Macgillivray AJ, Wood DR. The heterogeneity of mouse-chromatin nonhistone proteins as evidenced by two-dimensional polyacrylamide-gel electrophoresis and ion-exchange chromatography. Eur J Biochem. 1974;41(1):181–90.
4. O'Farrell PH. High resolution two-dimensional electrophoresis of proteins. J Biol Chem. 1975;250(10):4007–21.
5. Rabilloud T, Chevallet M, Luche S, Lelong C. Two-dimensional gel electrophoresis in proteomics: past, present and future. J Proteomics. 2010;73(11):2064–77.
6. Berth M, Moser FM, Kolbe M, Bernhardt J. The state of the art in the analysis of two-dimensional gel electrophoresis images. Appl Microbiol Biotechnol. 2007;76(6):1223–43.
7. The creative commons copyright licenses. https://creativecommons.org/licenses/.
8. Zhu H, Snyder M. Protein chip technology. Curr Opin Chem Biol. 2003;7(1):55–63.
9. Romanov V, Davidoff SN, Miles AR, Grainger DW, Gale BK, Brooks BD. A critical comparison of protein microarray fabrication technologies. Analyst. 2014;139(6):1303–26.
10. Gupta S, Manubhai KP, Kulkarni V, Srivastava S. An overview of innovations and industrial solutions in protein microarray technology. Proteomics. 2016;16(8):1297–308.
11. Pla-Roca M, Leulmi RF, Tourekhanova S, Bergeron S, Laforte V, Moreau E, et al. Antibody colocalization microarray: a scalable technology for multiplex protein analysis in complex samples. Molecular Cellular Proteomics. 2012;11(4):011460.
12. Rubina AY, Kolchinsky A, Makarov AA, Zasedatelev AS. Why 3-D? Gel-based microarrays in proteomics. Proteomics. 2008;8(4):817–31.
13. Yu X, Xu D, Cheng Q. Label-free detection methods for protein microarrays. Proteomics. 2006;6(20):5493–503.
14. Karas M, Bachmann D, Bahr U, Hillenkamp F. Matrix-assisted ultraviolet laser desorption of non-volatile compounds. Int J Mass Spectrom Ion Process. 1987;78:53–68.
15. Tanaka K, Waki H, Ido Y, Akita S, Yoshida Y, Yoshida T, et al. Protein and polymer analyses up to m/z 100 000 by laser ionization time-of-flight mass spectrometry. Rapid Commun Mass Spectrom. 1988;2(8):151–3.
16. Hou T-Y, Chiang-Ni C, Teng S-H. Current status of MALDI-TOF mass spectrometry in clinical microbiology. J Food Drug Anal. 2019;27(2):404–14.
17. Croxatto A, Prod'hom G, Greub G. Applications of MALDI-TOF mass spectrometry in clinical diagnostic microbiology. FEMS Microbiol Rev. 2012;36(2):380–407.
18. Hutchens TW, Yip T-T. New desorption strategies for the mass spectrometric analysis of macromolecules. Rapid Commun Mass Spectrom. 1993;7(7):576–80.

19. Wright GL Jr. SELDI proteinchip MS: a platform for biomarker discovery and cancer diagnosis. Expert Rev Mol Diagn. 2002;2(6):549–63.

20. Lambacher A, Jenkner M, Merz M, Eversmann B, Kaul RA, Hofmann F, et al. Electrical imaging of neuronal activity by multi-transistor-array (MTA) recording at 7.8 μm resolution. Appl Phys. 2004;79(7):1607–11.

21. Reddy G, Dalmasso EA. SELDI ProteinChip® array technology: protein-based predictive medicine and drug discovery applications. J Biomed Biotechnol. 2003;2003:874759.

22. Merchant M, Weinberger SR. Recent advancements in surface-enhanced laser desorption/ionization-time of flight-mass spectrometry. Electrophoresis. 2000;21(6):1164–77.

23. Issaq HJ, Veenstra TD, Conrads TP, Felschow D. The SELDI-TOF MS approach to proteomics: protein profiling and biomarker identification. Biochem Biophys Res Commun. 2002;292 (3):587–92.

24. Ndao M, Spithill TW, Caffrey R, Li H, Podust VN, Perichon R, et al. Identification of novel diagnostic serum biomarkers for chagas; disease in asymptomatic subjects by mass spectrometric profiling. J Clin Microbiol. 2010;48(4):1139.

25. Wilson LL, Tran L, Morton DL, Hoon DS. Detection of differentially expressed proteins in early-stage melanoma patients using SELDI-TOF mass spectrometry. Ann N Y Acad Sci. 2004;1022:317–22.

26. Engwegen JY, Gast MC, Schellens JH, Beijnen JH. Clinical proteomics: searching for better tumour markers with SELDI-TOF mass spectrometry. Trends Pharmacol Sci. 2006;27(5):251–9.

27. Ndao M, Rainczuk A, Rioux M-C, Spithill TW, Ward BJ. Is SELDI-TOF a valid tool for diagnostic biomarkers? Trends Parasitol. 2010;26(12):561–7.

28. Ruggeri FS, Šneideris T, Vendruscolo M, Knowles TPJ. Atomic force microscopy for single molecule characterisation of protein aggregation. Arch Biochem Biophys. 2019;664:134–48.

29. Deng X, Xiong F, Li X, Xiang B, Li Z, Wu X, et al. Application of atomic force microscopy in cancer research. J Nanobiotechnol. 2018;16(1):102.

30. Lynch M, Mosher C, Huff J, Nettikadan S, Johnson J, Henderson E. Functional protein nanoarrays for biomarker profiling. Proteomics. 2004;4(6):1695–702.

31. Krieg M, Fläschner G, Alsteens D, Gaub BM, Roos WH, Wuite GJL, et al. Atomic force microscopy-based mechanobiology. Nature Rev Phys. 2019;1(1):41–57.

32. Syu G-D, Dunn J, Zhu H. Developments and applications of functional protein microarrays. Mol Cell Proteomics. 2020;19(6):916–27.

33. Ayton LN, Barnes N, Dagnelie G, Fujikado T, Goetz G, Hornig R, et al. An update on retinal prostheses. Clin Neurophysiol. 2020;131(6):1383–98.

34. Sakmann B, Neher E. Patch clamp techniques for studying ionic channels in excitable membranes. Annu Rev Physiol. 1984;46:455–72.

35. Neher E, Sakmann B. Single-channel currents recorded from membrane of denervated frog muscle fibres. Nature. 1976;260(5554):799–802.

36. Martinez D, Py C, Denhoff MW, Martina M, Monette R, Comas T, et al. High-fidelity patch-clamp recordings from neurons cultured on a polymer microchip. Biomed Microdevices. 2010;12 (6):977–85.

37. Priest BT, McDermott JS. Cardiac ion channels. Channels. 2015;9(6):352–9.

38. Goldman DE. Potential, impedance, and rectification in membranes. J Gen Physiol. 1943;27 (1):37–60.

39. Hodgkin AL, Katz B. The effect of sodium ions on the electrical activity of giant axon of the squid. J Physiol. 1949;108(1):37–77.

40. Nattel S, Carlsson L. Innovative approaches to anti-arrhythmic drug therapy. Nat Rev Drug Discov. 2006;5(12):1034–49.

41. Vandenberg JI, Perry MD, Perrin MJ, Mann SA, Ke Y, Hill AP. hERG K+ channels: structure, function, and clinical significance. Physiol Rev. 2012;92(3):1393–478.

42. Zrenner E, Stett A, Rubow R. The challenge to meet the expectations of patients, ophthalmologists and public health care systems with current retinal prostheses. Invest Ophthalmol Vis Sci. 2020;61(7):2214.

43. Ridder BJ, Leishman DJ, Bridgland-Taylor M, Samieegohar M, Han X, Wu WW, et al. A systematic strategy for estimating hERG block potency and its implications in a new cardiac safety paradigm. Toxicol Appl Pharmacol. 2020;394:114961.

44. Danker T, Möller C. Early identification of hERG liability in drug discovery programs by automated patch clamp. Front Pharmacol. 2014;5:203.

45. Mahfooz K, Ellender TJ. Combining whole-cell patch-clamp recordings with single-cell RNA sequencing. In: Dallas M, Bell D, editors. Patch clamp electrophysiology: methods and protocols. New York: Springer; 2021. p. 179–89.

46. Zhou Y, Li H, Xiao Z. In vivo patch-clamp studies. In: Dallas M, Bell D, editors. Patch clamp electrophysiology: methods and protocols. New York: Springer; 2021. p. 259–71.

47. Brueggemann A, George M, Klau M, Beckler M, Steindl J, Behrends JC, et al. Ion channel drug discovery and research: the automated Nano-patch-clamp technology. Curr Drug Discov Technol. 2004;1(1):91–6.

48. Lewis PM, Rosenfeld JV. Electrical stimulation of the brain and the development of cortical visual prostheses: an historical perspective. Brain Res. 2016;1630:208–24.

49. Utah electrode array. https://www.blackrockmicro.com/electrode-types/utah-array/.

50. Maynard EM, Nordhausen CT, Normann RA. The Utah Intracortical electrode Array: a recording structure for potential brain-computer interfaces. Electroencephalogr Clin Neurophysiol. 1997;102(3):228–39.

51. Brüggemann A, Farre C, Haarmann C, Haythornthwaite A, Kreir M, Stoelzle S, et al. Planar patch clamp: advances in electrophysiology. In: Lippiat JD, editor. Potassium channels: methods and protocols. Totowa, NJ: Humana Press; 2009. p. 165–76.

52. Obergrussberger A, Friis S, Brüggemann A, Fertig N. Automated patch clamp in drug discovery: major breakthroughs and innovation in the last decade. Expert Opin Drug Discovery. 2021;16 (1):1–5.

53. Fernández E, Alfaro A, González-López P. Toward long-term communication with the brain in the blind by intracortical stimulation: challenges and future prospects. Front Neurosci. 2020;14:681.

54. Troyk PR. The intracortical visual prosthesis project. In: Gabel VP, editor. Artificial vision: a clinical guide. Cham: Springer; 2017. p. 203–14.

55. Rosholm KR, Boddum K, Lindquist A. Perforated whole-cell recordings in automated patch clamp electrophysiology. In: Dallas M, Bell D, editors. Patch clamp electrophysiology: methods and protocols. New York: Springer; 2021. p. 93–108.

Microelectrode Arrays, Implants, and Organs-on-a-Chip

11

Contents

Keywords

Microelectrodes · Action potential · Neural interfaces · Function recovery · Organ emulation

What You Will Learn in This Chapter

In this chapter, we will focus on Micro Electrode Arrays (MEA) for measuring and stimulating the electric activity of electroactive cells and tissues. You will learn about the development of neuronal networks and the multi-transistor arrays for high-resolution recordings. We will further cover the neural implants to restore hearing and vision. Finally, we will deal with the emerging field of organs-on-a-chip for emulating organs or even biological systems. You will be able to describe these technologies and make statements about their applications.

© The Author(s), under exclusive license to Springer Nature Switzerland AG 2021   291
A. Pasquarelli, *Biosensors and Biochips*, Learning Materials in Biosciences,
https://doi.org/10.1007/978-3-030-76469-2_11

11.1 Microelectrode Array (MEA)

A microelectrodes array (MEA) is a passive device, able to record bioelectric signals and deliver stimuli. In MEAs, several thin-film microelectrodes with a diameter of 5–30 µm are arranged on a small area (typically 0.5–5 mm^2). These are multiple independent sites for stimulating and recording the activity of electroactive cells. The electrodes fabrication requires specific technologies for providing a large effective surface leading to a low-noise level, without compromising the spatial resolution. Fractal titanium nitride (TiN) and iridium oxide (IrOx) are among the best materials for these applications due to the large specific capacity and the consequent charge injection limits [1]; beside them, metals like gold, platinum, tantalum, tungsten, and steel, or even silicon and the related CMOS technology are also in use. Fractal TiN microelectrodes with a diameter of 10 µm show a typical noise level of 15 µV p-p. More recently also PEDOT microelectrodes demonstrated excellent performances in terms of low noise and high sensitivity, especially in combination with carbon nanotubes [2]. Alternatively, indium tin oxide (ITO) or boron-doped diamond (BDD) can be used for transparent electrodes [3, 4]; the latter is also particularly suited for amperometric applications, like the detection of neurotransmitter release [5, 6].

There are two fundamental classes of MEAs: planar chips for in vitro applications investigating seeded cells as well as neural, muscle, or cardiac tissues [7–9] and shank probes for in vivo recordings of cortical or deep-brain activity [10]. In the following, we focus on planar chips only (Fig. 11.1).

As seen in Chap. 10, cell electrophysiology is based on gated ion channels. MEAs are able to detect the associated potential transients. Figure 11.2 shows in detail a neuronal action potential (AP), which is the basic signal unit adopted for communication within the

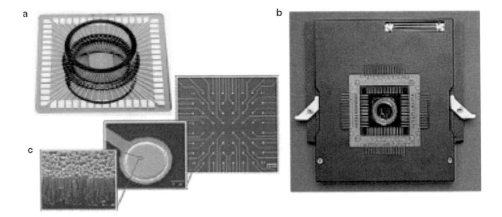

Fig. 11.1 (**a**) A microelectrode array (MEA) with attached culture dish; (**b**) MEA1060 amplifier with 60 channels for upright microscopes; and (**c**) a view of the MEA electrodes at different magnifications, revealing the fractal structure of the actual titanium nitride recording electrode (Reprinted from [9], © 2012, with permission from Springer Nature)

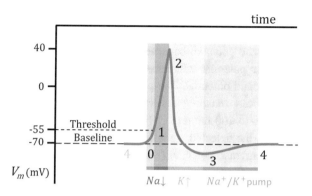

Fig. 11.2 Neuronal action potential: An external stimulus causes leading Na channels to open (0). [Na]out is larger than [Na]in, therefore Na$^+$ ions diffuse inward. After passing the threshold many more Na$^+$ channels open (1). Sodium is positive, so the cell becomes more positive and depolarizes. Soon after K$^+$ channels open (2), K$^+$ ions diffuse outward while Na$^+$ channels close. The membrane potential decreases passing the baseline (hyperpolarization) due to the slow closing of K$^+$ channels. Finally, the ion pumps restore the resting concentrations (3) and the transmembrane potential returns to the baseline (4)

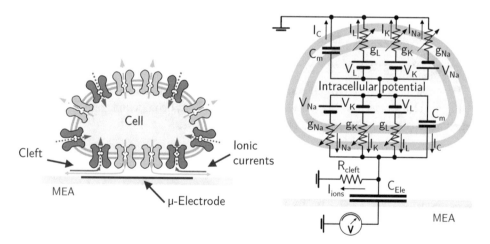

Fig. 11.3 (Left) Cartoon showing the cell–electrode interface. (Right) Equivalent circuit based on the Hodgkin–Huxley model. The currents flowing at the upper side of the cell face the bulk electrolyte at ground potential, while the ion flows in the thin cleft between the cell and the planar microelectrode produce voltage drops across the cleft resistance R_{cleft}. Such voltage transients are capacitively detected by the electrode and recorded by the active electronics outside the chip

nervous system. The AP is the result of a "tandem work" of Na$^+$ and K$^+$ channels, supported in the background by ion pumps. The neuron AP can be calculated with the Goldman–Hodgkin–Katz voltage equation (see paragraph 10.8) and represented in an equivalent circuit with the Hodgkin–Huxley model [11, 12] (Fig. 11.3 right).

Once seeded onto a MEA, the cells can communicate directly with microelectronic devices, but how does it work? The patch-clamp technique (see Chap. 10) allows to measure the transmembrane potential because the pipette, after penetration into the cell, monitors the intracellular potential while the extracellular solution is connected to the reference electrode at ground potential. The planar electrodes of a MEA do not penetrate into the cells, but they can sense capacitively the voltage transient related to action potentials, through the 30 to 100 nm thin cleft between the cell and the electrode. Due to this capacitive coupling, the detected signals are in a good approximation the first time derivative of the transmembrane potential (Fig. 11.3).

This planar technology allows noninvasive measurement with multisite recording; it is therefore ideally suited for investigations in the long term. An interesting example is represented by the study of how neuroactive cells grow their connections (neurites, synapses, axons, and dendrites) thus forming a neuronal network, and how the communication among them is progressively developed and optimized.

Figures 11.4 and 11.5 illustrate such a study carried out on a MEA. Fluorescence staining shows the progress in developing the anatomical structures of the network, while electrical recordings depict the development of spontaneous activity during the formation of a neuronal network. After 15 days in vitro (DIV) only few action potentials (spikes) are recorded by two electrodes only. One week later a few more events are observed, but the activity is still largely asynchronous. In the fourth week, synchronous oscillating bursts appear whose frequency further increases in the following 2 weeks.

Question 1

How can an electroactive cell communicate with a planar microelectrode?

Figure 11.6 shows the effect of various pharmacological compounds on the communication activity within the neural network described above.

11.2   CMOS-MEA

The next technical development stage involves integrated electronics. Electrodes on "classical" MEAs are passive elements that are fabricated at low density (up to ~250) on a chip. All these electrodes are electrically connected to the off-chip active electronics by means of planar wires patterned on the top surface of the chip. This technology allows up to a few hundred connections at most. For higher densities, it is necessary to adopt active microelectronics, for example, CMOS-based technologies. In this way it is possible to stack vertically the electrodes with the related switches and transistors of the front-end stage, which allow array densities of thousands of electrodes or even more [14–16]. Signal multiplexers, amplifiers, and converters are monolithically integrated on chip as well, thus drastically reducing the total pin count of the packaged chips. Figure 11.7 shows the schematic structure of an integrated CMOS-MEA for neural stimulation and recording, while an application example of this chip is depicted in Fig. 11.8.

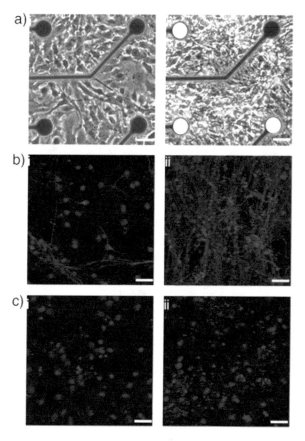

Fig. 11.4 Embryonic stem cell–derived neurons at different time points after initiating the differentiation. (**a**) Phase-contrast images of cells on the MEA. (**b, c**) Fluorescence images at 7 (i) or 28 days (ii) respectively after initiating the differentiation showing in blue the cell nuclei stained with DAPI (4′,6-Diamidino-2-phenylindole·2HCl), which binds selectively to the DNA. Immunocytochemical staining of β-tubulin (**b**) highlights neurites formation and of synaptophysin (**c**) highlights the formation of synapses. Note the dense network of neurites at 28 days after initiating the differentiation in comparison to 7 days (a, i, and ii) correlating to increased numbers of β-tubulin+ or synaptophysin + structures (b, c, i, and ii). White-labeled electrodes indicate the appearance of spontaneous action potentials while the black-labeled electrode detected no action potentials (a, i, and ii). Scale bar = 30 μm (Reprinted from [13], © 2007, with permission from Elsevier)

Beside the neuronal networks, CMOS-MEAs can deliver both temporal and spatial high-resolution information during experiments with just a few cells, out of which the individual behavior of a single cell can be extracted as shown in Fig. 11.9.

Question 2

What are MEAs? What are the differences between "classical" and CMOS MEAs?

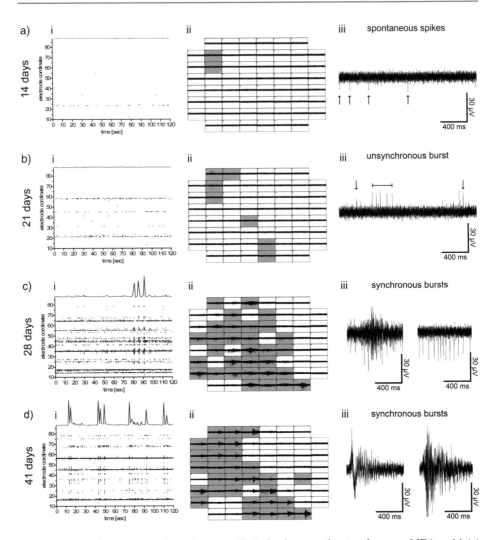

Fig. 11.5 Development of embryonic stem cell–derived neuronal network over a MEA at 14 (**a**), 21 (**b**), 28 (**c**), and 41 (**d**) days after initiating the differentiation of ES cell–derived neural precursors. (i) Spike raster plots (SPR) with frequency counts (panel above the SRPs). (ii and iii) Snapshot images of MEA recordings (Reprinted from [13], © 2007, with permission from Elsevier)

11.3   Implants: Neural Interfaces and Prostheses

Derived from MEAs, implants help patients to compensate for certain impairments. Applications include, among others, auditory prostheses, deep brain stimulation (Parkinson's disease, depression), spinal cord stimulation, visual restoration, functional neuromuscular/electrical stimulation (stand up and walk) bladder and bowel function restoration, and brain–machine interfaces [18, 19]. Currently, a limited impact can be

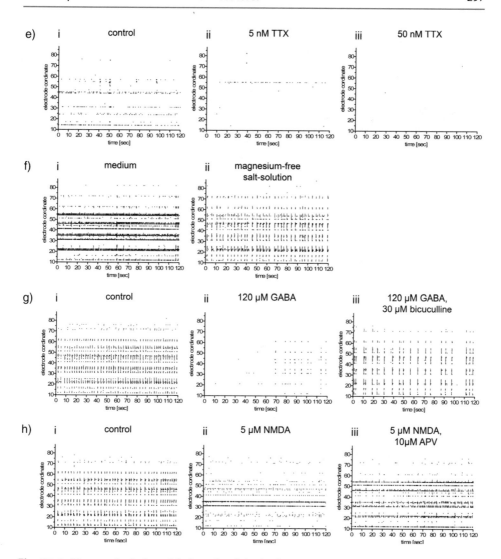

Fig. 11.6 Pharmacological modulation of activity in the same network described in the previous figure at 28 days after the induction of differentiation: (**a**) Burst activity reduced in a dose-dependent manner after application of 5 nM or 50 nM of sodium channel blocker TTX (tetrodotoxin). (**b**) The substitution of medium by a magnesium-free salt solution increases the synchronous burst activity and decreases uncorrelated activity. (**c**) Application of 120 µM of synaptic-acting agonists gamma-aminobutyric acid (GABA) to the magnesium-free salt solution reduces both synchronous and uncorrelated activity. Effect partly recovered by adding 30 µM bicuculline, a GABA-receptor antagonist. (**d**) Application of 5 µM N-methyl-D-aspartate (NMDA), another synaptic-acting agonists, results in a broad loss of synchronous burst activity while uncorrelated activity is preserved. The latter effect is partly antagonized by adding 10 µM of the antagonist 2-amino-5-phosphonovalerate (APV) (Reprinted from [13], © 2007, with permission from Elsevier)

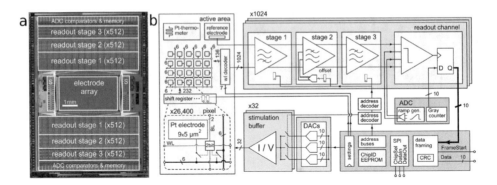

Fig. 11.7 Architecture of a rapidly reconfigurable CMOS-MEA: (**a**) Micrograph of the device with 26,400 microelectrodes arranged on an overall sensing area of 3.85×2.10 mm^2 at a pitch of 17.5 µm. An arbitrary subset of electrodes can be simultaneously connected to 1024 low-noise readout channels as well as 32 stimulation units (S). Each electrode or electrode subset can be used to electrically stimulate or record the signals of virtually any neuron on the array. (**b**) Block diagram of the implemented circuitry (Reprinted from [15], © 2015 by the Authors, CC-BY 3.0 [17])

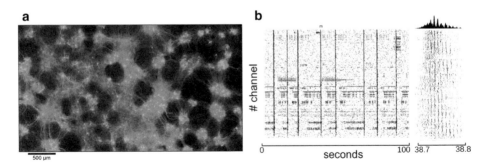

Fig. 11.8 (**a**) Fluorescence image of a network of cortical neurons grown over the CMOS-MEA described above. (**b**) Raster plot of 100 seconds of activity for all 1024 recording channels. The red marker indicates the time period between 38.7 and 38.8 seconds shown in the close up view to the right. The histogram at the top shows the number of spikes per time bin (Adapted from [15], © 2015 by the Authors, CC-BY 3.0 [17])

found at the neural interface and close attention to the integration between electrodes and tissue should improve the functional yield and the lifetime of the implants. Biocompatibility of the adopted technologies and bio-functionalization of the electro-active areas within a device appear to be crucial for solving this problem.

11.3.1 Deep Brain Stimulation (DBS)

Deep Brain Stimulation (DBS) uses a surgically implanted pacemaker to deliver electrical stimulation to those regions in the deep brain that control movement, thus blocking the

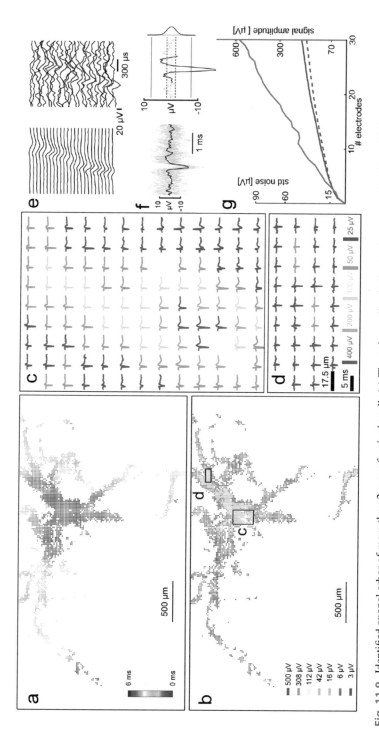

Fig. 11.9 Identified axonal arbor of more than 2 mm of a single cell. (**a**) The color coding indicates the propagation time. (**b**) Same neuron, electrodes show the signal amplitude. (**c**) and (**d**) averages (30 to 50 avg) from two areas indicated in (**b**). (**e**) Left: 25 traces from electrodes that detect axonal signals, averaging (50 APs). Right: single axonal AP in the noise. Red dots indicate the timing of the negative peak. (**f**) Left: 25 APs overlaid and aligned. Right: spatial averaging of traces aligned; green lines indicate 4.5 standard deviations, the detection threshold. (**g**) The signal vs. noise scales with a square root (Reprinted from [15]. © 2015 by the Authors, CC-BY 3.0 [17])

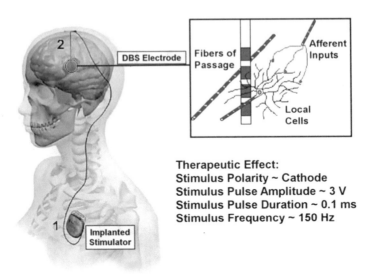

Fig. 11.10 Schematic representation of the DBS system (Adapted from [27], © 2015, with permission from Springer Nature)

abnormal neural activity that causes Parkinson's disease (PD) and dystonia, symptoms such as tremor, rigidity, slowed movement, and walking problems [20, 21]. This technique is providing since more than three decades significant improvements of life quality to affected patients with intractable tremor or complications related to the therapy based on levodopa. However, there is still a significant headroom for improving the clinical results, especially in terms of anatomical target localization during surgery and programming individualized stimulations [22, 23]. More recently, DBS was proposed in other clinical applications like the neurorehabilitation for stroke and for treating psychiatric disorders [24–26]. A DBS system consists of three components (Fig. 11.10):

1. The electrodes: inserted through a small opening in the skull and implanted in the targeted area of the brain.
2. The wires passed under the skin of the head, neck, and shoulder, connecting the electrodes to the pacemaker.
3. The pacemaker, usually implanted under the skin near the collarbone .

11.3.2 Cochlear Implant

Another example for multielectrode-based prosthesis is the cochlear implant for hearing restoration in profoundly deaf persons (Fig. 11.11), including children with congenital hearing loss, having an intact auditory innervation [28]. This implant is a mature technology; in fact by 2012, more than 320,000 prostheses had been implanted worldwide.

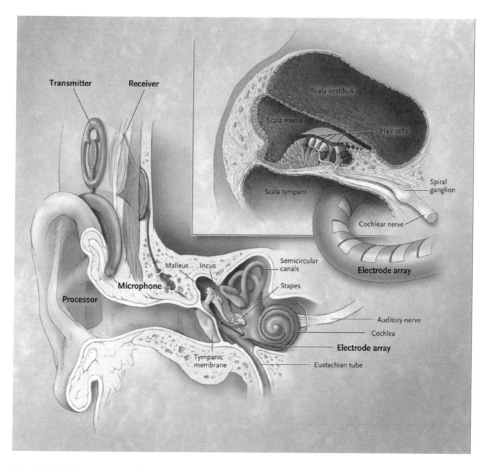

Fig. 11.11 Components of a cochlear implant related to the anatomy of the ear (Reprinted from [30], © 2020, with permission from Elsevier)

However, since these implants rely on audio processors and related algorithms, both hardware and software are progressing further [29–31].

Remarkably, the cochlear implant is indicated at all ages; for example, in the United States about 40% of the devices have been implanted in children [32]. This availability even for infants under 12 months has a decisive impact on the chances to achieve normal levels of spoken language [33].

Cochlear implants work similarly to a spectrum analyzer: A microphone placed at the external ear converts the environmental sound into an electrical audio signal, which is then decomposed into several spectral bands (e.g., 22) by the processor. These spectral audio components are wirelessly routed to the linear electrode array implanted in the cochlea within the *scala timpani*, in such a way that every electrode interfaces the specific spectral region of the cochlea, thereby stimulating the auditory nerve at the frequencies corresponding to the local sensitivity.

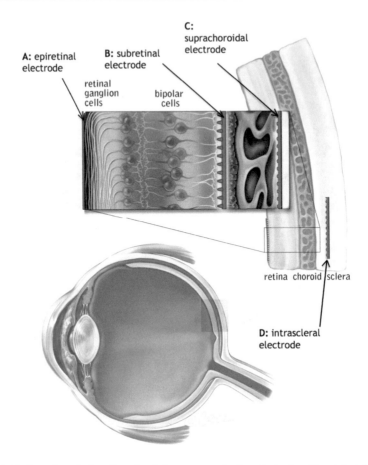

Fig. 11.12 Options for placing the stimulating array of microelectrodes (Reprinted from [40], © 2014 by the Authors, CC-BY 4.0 [17])

Question 3

What is a DBS system used for and what are its components?

11.3.3 Vision Restoration Implants

In case of loss of eyesight (e.g., macular degeneration, retinitis pigmentosa), implants can be used to partially restore vision [34–37]. A photo-sensor array measures the light signals and an array of microelectrodes stimulates the optic nerve terminations or alternatively the occipital visual cortex [38, 39]. Three approaches have been proposed for the retinal stimulation (Fig. 11.12): subretinal [41–44], epiretinal [45–47] and suprachoroidal [40, 48].

The light sensor array of both the epiretinal and suprachoroidal implants is an external head-mounted camera connected via a transdermal cable [40, 45] or wirelessly [47] to the stimulation array, while in subretinal implants the sensing array (photodiodes) and the stimulating array (microelectrodes) are integrated pixel by pixel in the same chip [44, 50] or wirelessly via near-infrared (NIR) projection [43]. Table 11.1 shows a partial overview; more information is available in [37, 51].

The development of these prostheses is not as advanced as in the case of cochlear implants; however, three products, namely the Alpha AMS from Retina Implant (Germany), the Argus® II from Second Sight Medical Products (USA), and the IRIS® II from Pixium Vision (France), gained market approval in several countries. In the following, we focus on the subretinal approach adopted by Retina Implant AG.

The development of implants requires several studies for a full understanding of how the device–tissue interface can work and other related issues like biocompatibility and long-term reliability. To this purpose, it is common to adopt suitable animal models. In the specific case of the retina implant, electrical retinal stimulation were studied ex vivo on a MEA after explantation of the retina from a chicken or rat eye [52–54].

For a better understanding, we briefly recall the image conversion process in the eye. This can be described in terms of a multistep process:

- Two-dimensional, optical image is projected onto the retina.
- Light is absorbed in the photoreceptors.
- Electric activity occurs in retinal cells.
- Information is processed in the neural network.
- Coded information is retinotopically corrected and projected in the visual cortex.
- Basic information is perceived in terms of brightness/darkness and localization.

From the mentioned studies on animal models, the developing team predicted that the electrical image stimulation of the retina works better from the backside (subretinal). In this way, the spatial arrangement of the electrodes matches that of the photoreceptors. In contrast, an undistorted image transmission at the front side of the retina is more difficult, due to the bundled nerve fibers (Fig. 11.13).

Retinal neurons can be successfully stimulated in patients with severe retinal degeneration, as long as the optic nerves are intact. Electric subretinal stimulation yields an ordered activation of the retina, which elicits retinotopic neural activity in the visual cortex. With the proposed implant (Fig. 11.14) the patients should expect a correct, spatially ordered perception with a spatial resolution of 0.5°–1° (Fig. 11.15).

Despite of all efforts over more than 20 years, it was not possible to meet the expectation of more than 40%–50% of the total pool of candidates and volunteers who received a retinal implant, that is, 72 persons with ALPHA AMS and more than 350 persons with ARGUS II). Moreover, considering the complicated regulatory procedures and the troublesome coverage of costs in the health insurance systems of many countries, the manufacturing of both ALPHA AMS and ARGUS II was discontinued in 2019 [56].

Table 11.1 Overview of implants for restoring vision (Reprinted from [49], © 2020, with permission from Springer Nature)

Device	Type	How it works:	Benefits	Limitations
ARGUS II Source: Courtesy of SecondSight	Epiretinal implant with external camera	External camera records and transmits pulses wirelessly through microelectrode array to inner retina	Unaffected by corneal or lens opacities due to wireless transmission from camera to array; long-term safety demonstrated	Visual perception through fixed camera independent of eye movement; 60 electrodes limits potential visual acuity
IRIS II Source: Courtesy of Pixium Vision	Epiretinal implant with external camera	External camera records and transmits pulses wirelessly through microelectrode array to inner retina	Exchangeable system allows for replacement; unaffected by lens or corneal opacities due to wireless transmission from camera to array	Visual perception through fixed camera is independent of eye movement; 150 electrodes but requires further studies to demonstrate efficacy

ALPHA IMS/AMS	Subretinal implant	Photodiode array in subretinal space that detects light and converts it to electrical signals transmitted to overlying retina	Vision generated corresponds with natural eye movement due to subretinal implant itself directly detecting light	Affected by opacities between cornea and retina due to requiring light to reach the retina; 1500 (IMS) or 1600 (AMS) electrodes but visual acuity significantly below potential
PRIMA	Subretinal implant with external camera	External camera records video that is processed and projected via infrared laser patterns through the pupil to subretinal implant, where photovoltaic array converts the light to electrical signals stimulating bipolar cells	Design of subretinal implant includes ground grid around each stimulating electrode allowing theoretical improvement of resolution	Affected by opacities between cornea and retina due to requiring light to reach the retina; 378 pixels (electrodes) but requires clinical trials to demonstrate efficacy

Source: Courtesy of Pixium Vision

(continued)

Table 11.1 (continued)

Device	Type	How it works:	Benefits	Limitations
Bionic Vison Australia	Suprachoroidal implant with external camera	External camera records and transmits pulses wirelessly through microelectrode array through choroid to inner retina	Unaffected by corneal or lens opacities due to wireless transmission from camera to array	Visual perception through fixed camera is independent of eye movement; 44 channels (electrodes) limits potential visual acuity and requires clinical trials to demonstrate efficacy
Orion Cortical Implant	Cortical implant with external camera	External camera records and transmits pulses wirelessly to microelectrode array to medial occipital lobe	Unaffected by corneal or lens opacities due to wireless transmission from camera to array; potential for use in patients with significant inner retina and/or optic nerve degeneration	Visual perception through fixed camera is independent of eye movement; 60 electrodes limits potential visual acuity; requires clinical trials to demonstrate efficacy and long-term safety

Source: Courtesy of Ayton et al. 2014, originally from Bionic Vision Technologies

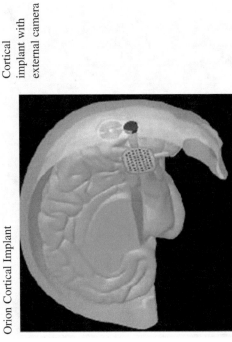

Source: Courtesy of SecondSight

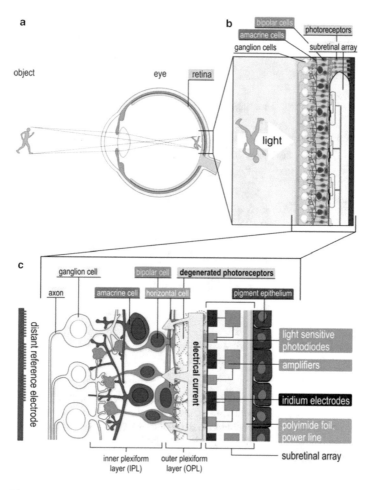

Fig. 11.13 The subretinal approach. (**a**) The image is projected through the lens onto the subretinal implant (**b**) with its light sensitive micro-photodiode array. (**c**) The signals from the photodiodes, amplified point by point are sent to the electrodes, which directly stimulate the bipolar cells, resulting in neuronal processing of the signals by the retinal bipolar cells, amacrine cells, and ganglion cells. Ganglion cells send the processed information via their axons to the brain. In contrast, epiretinal stimulation does not use the inner retina processing by means of the various synaptic connections but preprocesses the image in an external computer (Reprinted from [55], © 2017, with permission from Springer Nature)

At present, research on visual restoration implants with retinal stimulation and the related clinical trials are continuing (e.g., Pixium and Bionic Vision Technologies). Alternatively, the electric stimulation of the visual cortex, after its introduction in the late 1960s and implantation in patients in the 1970s while remaining an experimental approach, is presently regaining much attention [38, 57, 58]. This renewed interest relies in its potential to overcome some limitations of the retinal approach, like degeneration or damage of the optic nerve, advanced degeneration of the retina, glaucoma or optic atrophy, and

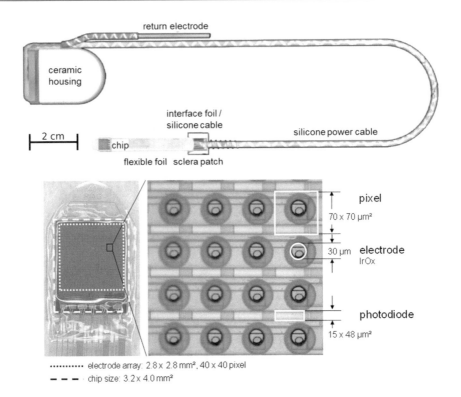

Fig. 11.14 Technological implementation of the subretinal implant "Alpha AMS" by Retina Implant AG. (Top) The whole assembly, i.e., intraocular chip, connection cable, and electronics in a hermetic housing to be implanted behind the ear. (Bottom) Detail view of the light sensitive chip, containing 1600 pixels. The electrodes are fabricated with iridium oxide and the passivation is out of polyimide (Reprinted by permission from [50], © MYU 2018)

Fig. 11.15 (Left) Patient looking at of Bridge of Sighs in Oxford. (Right) Sketch of the patient's perception while seeing with the Retina Implant Alpha IMS, reporting a blurred scene in several levels of gray consisting of an arch and neighboring buildings (Reprinted from [55], © 2017, with permission from Springer Nature)

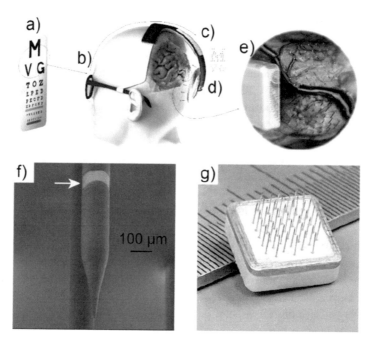

Fig. 11.16 Wirelessly powered and controlled cortical vision prosthesis system. (**a**) Objects captured via a camera (**b**) are processed (**c**) into instructions that are sent via a (**d**) wireless link to the (**e**) implanted device. (**f**) Individual polyimide insulated platinum–iridium electrode. An annular platinum ring is exposed on the electrode shaft (arrow). (**g**) Each implant consists of 43 penetrating electrodes with electronics encapsulated in a ceramic case (Reprinted from [63], © 2020, with permission from IOP Publishing)

technological issues like biocompatibility, implant delamination, and degradation of the protective encapsulation. Moreover, the surgical implantation of retinal prostheses is relatively difficult and requires a high degree of placement precision. Interfacing the visual cortex appears advantageous because the neurons in the visual regions are usually intact. The size of the primary visual cortex (~ 1 cm^3) allows placing close to the neurons a few hundred penetrating electrodes, for example, a Utah Electrode Array [59, 60], which are expected to provide enough visual acuity for facilitating daily living activities like low-speed reading and self-orientation in a nonfamiliar environment. However also this approach has to face issues, like the biotolerability, the reliable functionality in the long term and the risk of infections due to the directly wired feedthrough connections between the external electronics and the implanted electrodes. The best solution to the latter issue appears to be a wireless connection as proposed in [61–63] (Fig. 11.16).

11.4   Biocompatibility

One of the most challenging aspects in the development of neural implants is the biocompatibility of all materials in contact with the biological environment and in particular with tissues of the receiver's body. Biocompatibility is a concept frequently associated with prosthetic implants, and can be defined in this way: "Biocompatibility is the capability of a prosthesis implanted in the body to exist in harmony with tissue without causing deleterious changes" [64].

In case of limited biocompatibility, the body may respond with a variety of reactions in the short as well in the long term, including [64]:

- Protein adsorption and desorption characteristics
- Generalized cytotoxic effects
- Neutrophil activation
- Macrophage activation, foreign body giant cell production, granulation tissue formation
- Fibroblast behavior and fibrosis
- Microvascular changes
- Tissue−/organ-specific cell responses
- Activation of clotting cascade
- Platelet adhesion, activation, aggregation
- Complement activation
- Antibody production, immune cell responses
- Acute hypersensitivity/anaphylaxis
- Delayed hypersensitivity
- Mutagenic responses, genotoxicity
- Reproductive toxicity
- Tumor formation

From the implant point of view, the above reactions may lead on the one hand to delamination, degradation, and loss of materials, and on the other hand to induction of fibrosis with the consequent encapsulation of the implant. Both phenomena lead to loss of function, in the first case because the active material is lost, and in the second case which is particularly crucial for electrodes, because of the insulating effect, which hinders the delivery of the necessary electric stimuli. The conventional surfaces passivation materials adopted in microelectronics (SiO, SiN, etc.) fail in the long term when adopted for implants (Fig. 11.17). In the meantime, several polymers such as silicone rubbers, Teflon, polyimide, and parylene demonstrated suitability for biocompatible insulations [65]. Regarding the materials for the fabrication of electrodes, which cannot be insulated, good results have been reported for platinum–iridium alloy in case of standard electrodes [63, 66] and for iridium oxide in case of microelectrodes [1, 50]. More recently novel materials like graphene–platinum composites, conductive polymers, and amorphous silicon carbide have been investigated and look very promising for implantable neural

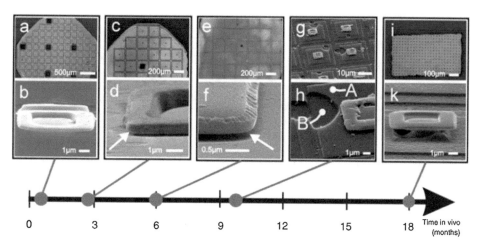

$$0 \qquad 3 \qquad 6 \qquad 9 \qquad 12 \qquad 15 \qquad 18 \quad \text{Time in vivo (months)}$$

Fig. 11.17 Tests run during the early stages of development of a retinal implant showing degradation of passivation which leads to loss of functionality of the micro-photodiode arrays (MPDA): (**a, b**) MPDA explanted after 3 weeks from a rabbit eye. No morphological damage was observed. (**c, d**) MPDA explanted from rabbit after 2.6 months. Light microscopy reveals laterally inhomogeneous interferences indicating initial corrosion of the passivation layer; the SEM micrograph below shows a gap between the rim of electrode and passivation layer indicated by the arrow. (**e, f**) MPDA explanted from micro-pig after 6 months. At the periphery of the chip, the passivation layer is almost completely lost. The pronounced interference is indicative for a strong lateral variation of the oxide thickness on a typical length scale of 20–100 mm. (**g, h**) MPDA explanted from rabbit after 10 months. The passivation layer is completely dissolved and pit corrosion to the underlying silicon is observed. The letters (A, B) in Fig. 11.6h indicate the location of EDX analysis. (**k, l**) MPDA explanted from rat after 18 months. Progressive corrosion of silicon is observed. In contrast, no sign of morphological damage is found on the electrodes indicating biostability of titanium nitride (Reprinted from [69], © 2002, with permission from Elsevier)

interfaces [67, 68] The expected functional lifetime of implants fabricated with appropriate materials should be of at least 10 years.

Question 4

Explain how retinal implants work.

What are the factors and challenges still limiting its application?

11.5 Organoids and Organs-on-a-Chip

In this last section, we give a prospective look at two emerging disciplines which combine the progresses in the fields of stem cells manipulation and self-organized 3D tissue cultures with microfluidics and biosensing technologies: The Organoids and their utilization as Organs-on-a-Chip.

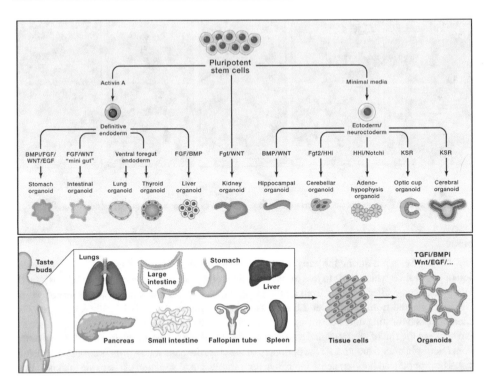

Fig. 11.18 (Top) Schematic of the various organoids that can be grown from PSCs and the employed growth conditions. (Bottom) Body source organs for culturing ASC-derived organoids (Reprinted from [71], © 2016, with permission from Elsevier)

An organoid can be defined as a 3D ensemble, containing a variety of organ-specific cell types, which develop and differentiate from stem cells or organ progenitors, self-organize themselves through cell-sorting and spatially restricted lineage commitment similarly to the organ to mimic, and exhibit some of the functions specific for that organ [70, 71].

Basically speaking, organoids can grow either form pluripotent stem cells (PSCs), that is, both embryonic stem cells [72] and induced pluripotent stem cells [73], or from organ-restricted adult stem cells (ASCs) [74, 75] (Fig. 11.18). Their differentiation and proliferation can be conveniently directed with tailored cocktails of growth factors, specific for leading to the formation of the desired organ [71]. However special attention is required to overcome issues like the variability of tissue self-organization, the limited range of function reproduction in comparison with the organ to mimic, and the limited culturing life span, just to mention a few. Addressing such problems requires the adoption of appropriate engineering methods [76, 77].

Nowadays, organoids are considered an extremely powerful tool for research in the fields of personalized, regenerative, and precision medicine. In fact, they preserve the genetic and phenotypic characteristics of the donor, while recreating models of the donor's

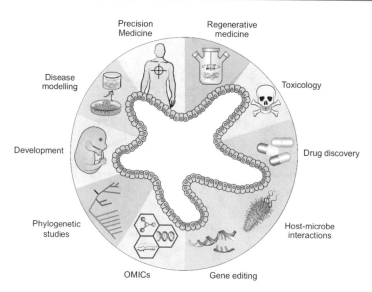

Fig. 11.19 Diverse applications of organoid technology (Reprinted from [81], © 2020, the American Physiological Society, CC-BY 4.0 [17])

organs, for investigating in vitro their development and related diseases, here including the genetic, infectious, and malignant ones [78]. Possible studies include, among others, basic physiology, genomic sequencing, gene editing, gene expression analysis, response to infectious pathogens, analyzing transplantation and tissue repair, drugs screening, drug development, and immunotherapy [78–81] (Fig. 11.19).

From a biosensing point of view, that is, the aim of this book, the most promising exploitation of organoids is represented by the so-called Organ-on-a-Chip (OoC) approach. The devices of this class merge organoids and, if appropriate, other 3D-bioenginered constructs, with microfluidic structures and microfabricated transduction technologies like electrodes, light sources and sensors, thermometers, sensors of chemical parameters, and even mechanical AFM-tips, for emulating, stimulating, and monitoring complex organ behaviors, while working in sub-milliliter volumes.

A wide variety of single OoC models was developed in the last decade (Fig. 11.20), reaching a high level of quality and robustness, and revolutionizing the field of in vitro functional testing for precision medicine [82–85]. Out of these experiences, a limitation of this approach became evident, because the organs in the body do not work alone, but their functions are interdependent, they communicate via a variety of mediators transported in the blood and lymph streams like hormones, exosomes, cells, and other signaling entities. In addition, many pathological conditions are well known for involving multiple organs.

This awareness is leading to the development of Multi-Organ-on-a-Chip and Body-on-a-Chip (also called Human-on-a-Chip) systems (Fig. 11.21), which are now able to emulate entire biological processes and perform multiparametric systemic studies, providing

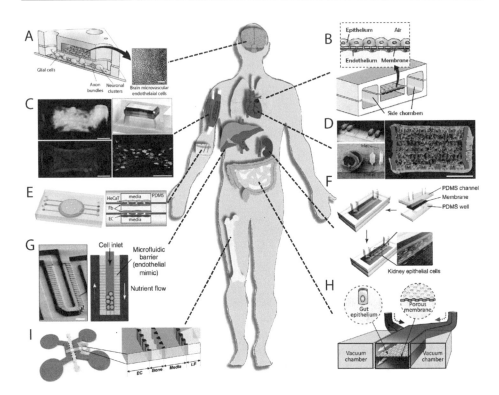

Fig. 11.20 Different OoC models and corresponding body organs: (**a**) Brain-on-a-chip platform demonstrating a layer of human neuronal and glial cells, which interact through a membrane containing perforations and a monolayer of human BMECs (scale bar, 250 μm). (**b**) Lung-on-a-chip mimicking lung microarchitecture and induced cyclic mechanical distortion of the alveolar–capillary interface. (**c**) Skeletal muscle microtissue and representative immunofluorescence imaging demonstrating skeletal muscle alignment (F-actin in red) (scale bars, 100 μm). (**d**) Thick AngioChip cardiac tissues placed beside a slice of an adult rat heart, and immunostaining of F-actin (green) of the cross-section of the thick multilayer human AngioChip cardiac tissue (scale bar, 1 mm). (**e**) Skin-on-a-chip device filled with fluid of three different colors comprising of three PDMS layers and two porous membranes, which allow for stacking of multiple cell types present in the skin. (**f**) Kidney-on-a-chip model that incorporates stacked PDMS layers with microchannels and a PDMS well separated by a porous polyester material, which allows 3D kidney microarchitecture. (**g**) Liver-on-a-chip mimicking hepatic microarchitecture containing (scale bar, 50 μm). (**h**) Gut-on-a-chip device containing a flexible porous ECM-coated membrane lined by gut epithelial cells in the middle and vacuum chambers on both sides to mimic intestinal peristalsis. (**i**) Bone-on-a-chip model consisting of four parallel channels separated by 100 μm gaps via microposts allowing for paracrine communication between ECs and stromal cells (connective tissue) during vessel formation (Reprinted from [85], © 2017, with permission from John Wiley and Sons)

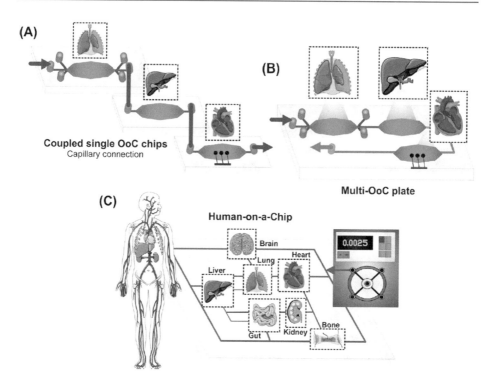

Fig. 11.21 Schematic representation of main approaches for developing Multi-OoC Systems. (**a**) Coupling multiple single-OoC devices, each modeling a different organ, via capillary connection or (**b**) a microfluidic motherboard. (**c**) Integration of different organ models in a single plate, an approach that is more in line with the body-on-a-chip philosophy (Reprinted from [86], © 2021 by the authors, CC-BY 4.0 [17])

comprehensive and accurate information not previously accessible with single-organ models [86].

Besides, multi-OoCs represent a very attractive alternative to the use of animal models in many studies [87], well supporting the 3Rs principle (replacement, reduction, and refinement).

We conclude this short perspective introduction with a final remark: OoC and especially Multi-OoC systems have the potential to deliver data on physiological and pathological processes, and from the other applications described above, with an unprecedented degree of accuracy and reliability, thus enabling precision personalized medicine and strongly reducing the need for animal models. However, reaching these goals will only be possible by integrating in such systems the power of high-tech resources spanning from biofabrication to artificial intelligence, from robotics to automation [88] and even handling of Big Data [89–91].

Question 5

Explain how scientific progress let to organoids and OoCs.

Mention a few fields in which multi-OoC can revolutionize medical care.

Answer 1

Electroactive cell possess gated ion channels that open and close during spontaneous or stimulated activities, thus letting ions flow across the cell membrane. This flow of mobile charges results in a transient voltage drop along the non-negligible resistance of the electrolyte present in the thin cleft between cell and electrode. The electrode senses such voltage transient capacitively; therefore, the detected signal is in good approximation the first time derivative of the transmembrane potential.

Answer 2

MEAs are planar chips able to detect electrophysiological signals from electroactive cells and tissues and deliver stimuli to them, consisting of several microelectrodes arranged on regular patterns (e.g., square or hexagonal grids), fabricated with the technologies of microelectronics.

Classical MEAs are fully passive devices, usually fabricated on glass substrates and having at the edges one pad per electrode, for contacting the off-chip active readout electronics.

CMOS-MEAs are integrated active electronic devices fabricated on silicon wafers. The presence of active analog and digital electronics allows the fabrication of arrays with a much larger number of electrodes (up to tens of thousands), thanks to the availability of amplifiers, multiplexers, and converters, which enormously simplify the pin-count at the communication interface. Due to the high density of electrodes, CMOS-MEAs map the bioelectric fields with a spatial resolution of a few micrometers.

Answer 3

The deep-brain stimulation device is a neuroprosthetic implant used for stimulating specific areas of the deep brain. The most prominent use is related to the Parkinson's disease for the prevention of tremor and allow a regular function of skeletal muscles. More recently the DBS approach was proposed also for selected psychiatric disorders.

A DBS system consists of three components:

1. The electrodes inserted through a small opening in the skull and implanted in the targeted area of the brain.
2. The wires passed under the skin of the head, neck, and shoulder, connecting the electrodes to the pacemaker.
3. The pacemaker, usually implanted under the skin near the collarbone.

Answer 4

A retinal implant consists of an image sensor, a processing unit (analog or digital), and an array of microelectrodes. There are different architectures depending on the placement and properties of all three components. Some approaches use as image sensor an external camera mounted on a goggle, while other systems place a photodiode array behind the retina (subretinal implantation). The detected information can be elaborated by a digital processor and coupled to the optic nerve by means of an epiretinally implanted MEA or conditioned pixel by pixel, with analog electronics, and locally coupled at subretinal level via intercalated electrodes on the same chip, thus matching the projection of the original image on the retina. The communication between the active electronics and the MEA is either wired over cable, or wireless via radiofrequency or even via an infrared laser pattern.

Limiting factors are both clinical and technological issues like

- Difficulties of the surgical implantation.
- Degeneration or damage of the optic nerve.
- Advanced degeneration of the retina.
- Glaucoma or optic atrophy.
- Biocompatibility of technological materials.
- Implant delamination.
- Degradation of the protective encapsulation.

Answer 5

Organoids could be developed after the availability of pluripotent stem cells and adult stem cells. Scientist learned how to induce the desired differentiation of those stem cells in order to achieve the growth of 3D tissues mimicking in terms of cell composition, structure, and functions of the desired organ to model.

Merging the organoid technology with microfluidics and biosensor/biochip technologies allows nowadays to produce Organs-on-a-Chip as functional investigative miniaturized organ models, which preserve both genotype and phenotype of the donors.

Multi-OoC will permit the investigation of diseases and the development of personalized therapies in an unprecedented manner thanks to the precise modeling of the entire biological system involved in the pathology to treat. Such treatments may include among others tailored drugs, gene editing, and immunotherapies.

Moreover, multi-OoC will allow a decisive reduction of animal models in many studies, well supporting the 3Rs principle.

Take-Home Messages

- MEAs allow investigating the activity of multicellular ensembles, for example, neural networks and electroactive tissues, and their response to external chemical and physical stimuli.
- Electroactive cells communicate by means of action potentials and signaling chemicals like hormones and neurotransmitters.
- Potentiometric MEAs detect action potentials, while amperometric MEAs can detect the release of RedOx active molecules.
- CMOS-MEAs allow recording signals at a much higher spatial resolution.
- Neuroprosthetic implants are based on electrode arrays and can restore lost functions.
- Deep brain stimulation and cochlear implants are mature technologies widely adopted.
- Other prosthetic approaches like in case of vision restoration are still in a development stage.
- Organoids can replicate in a miniaturized scale the organs of a donor, making organ models available for accurate in vitro studies in a variety of applications.
- Organs-on-a-Chip merge Organoids with microfluidics and biosensing technologies for multivariate studies of physiological and pathological processes.
- Body-on-a-Chip platforms allow complete systemic analyses of complex processes involving cross-interactions of several organs.
- The potential of Multi-OoCs will be fully exploited by combining multiple high-tech resources including, among others, artificial intelligence, and robotics.

References

1. Cogan SF. Neural stimulation and recording electrodes. Annu Rev Biomed Eng. 2008;10:275–309.
2. Gerwig R, Fuchsberger K, Schroeppel B, Link G, Heusel G, Kraushaar U, et al. PEDOT–CNT composite microelectrodes for recording and electrostimulation applications: fabrication, morphology, and electrical properties. Frontiers in Neuroengineering. 2012;5:8.
3. Gross GW, Wen WY, Lin JW. Transparent indium-tin oxide electrode patterns for extracellular, multisite recording in neuronal cultures. J Neurosci Methods. 1985;15(3):243–52.
4. Pasquarelli A, Wang X. Enhancing the transparency of highly electroactive BNCD-MEAs. Physica Status Solidi. 2018;215:1800211.
5. Gosso S, Turturici M, Franchino C, Colombo E, Pasquarelli A, Carbone E, et al. Heterogeneous distribution of exocytotic microdomains in adrenal chromaffin cells resolved by high-density diamond ultra-microelectrode arrays. J Physiol. 2014;592(15):3215–30.
6. Carabelli V, Marcantoni A, Picollo F, Battiato A, Bernardi E, Pasquarelli A, et al. Planar diamond-based multiarrays to monitor neurotransmitter release and action potential firing: new perspectives in cellular neuroscience. ACS Chem Neurosci. 2017;8(2):252–64.

7. Nisch W, Böck J, Egert U, Hämmerle H, Mohr A. A thin film microelectrode array for monitoring extracellular neuronal activity in vitro. Biosens Bioelectron. 1994;9(9):737–41.

8. Stett A, Egert U, Guenther E, Hofmann F, Meyer T, Nisch W, et al. Biological application of microelectrode arrays in drug discovery and basic research. Anal Bioanal Chem. 2003;377 (3):486–95.

9. Meyer T, Boven K-H, Günther E, Fejtl M. Micro-electrode arrays in cardiac safety pharmacology. Drug Saf. 2004;27(11):763–72.

10. Steinmetz NA, Koch C, Harris KD, Carandini M. Challenges and opportunities for large-scale electrophysiology with neuropixels probes. Curr Opin Neurobiol. 2018;50:92–100.

11. Hodgkin AL, Huxley AF. A quantitative description of membrane current and its application to conduction and excitation in nerve. J Physiol. 1952;117(4):500–44.

12. Beeman D. Hodgkin-Huxley model. In: Jaeger D, Jung R, editors. Encyclopedia of computational neuroscience. New York: Springer; 2013. p. 1–13.

13. Illes S, Fleischer W, Siebler M, Hartung H-P, Dihné M. Development and pharmacological modulation of embryonic stem cell-derived neuronal network activity. Exp Neurol. 2007;207 (1):171–6.

14. Lambacher A, Jenkner M, Merz M, Eversmann B, Kaul RA, Hofmann F, et al. Electrical imaging of neuronal activity by multi-transistor-array (MTA) recording at 7.8 μm resolution. Applied Physics A. 2004;79(7):1607–11.

15. Müller J, Ballini M, Livi P, Chen Y, Radivojevic M, Shadmani A, et al. High-resolution CMOS MEA platform to study neurons at subcellular, cellular, and network levels. Lab Chip. 2015;15 (13):2767–80.

16. Ogi J, Kato Y, Matoba Y, Yamane C, Nagahata K, Nakashima Y, et al. Twenty-four-micrometer-pitch microelectrode array with 6912-channel readout at 12 kHz via highly scalable implementation for high-spatial-resolution mapping of action potentials. Biointerphases. 2017;12(5):05f402.

17. The creative commons copyright licenses. https://creativecommons.org/licenses/.

18. Kilgore KL, Anderson KD, Peckham PH. Neuroprosthesis for individuals with spinal cord injury. Neurol Res. 2020:1–13.

19. Bhadra N, Chae J. Implantable neuroprosthetic technology. NeuroRehabilitation. 2009;25:69–83.

20. Benabid AL, Pollak P, Louveau A, Henry S, de Rougemont J. Combined (thalamotomy and stimulation) stereotactic surgery of the VIM thalamic nucleus for bilateral Parkinson disease. Appl Neurophysiol. 1987;50(1–6):344–6.

21. Volkmann J. Deep brain stimulation for the treatment of Parkinson's disease. Journal of clinical neurophysiology. 2004;21(1):6–17.

22. Hartmann CJ, Fliegen S, Groiss SJ, Wojtecki L, Schnitzler A. An update on best practice of deep brain stimulation in Parkinson's disease. Ther Adv Neurol Disord. 2019;12:1756286419838096.

23. Sui Y, Tian Y, Ko WKD, Wang Z, Jia F, Horn A, et al. Deep brain stimulation initiative: toward innovative technology, new disease indications, and approaches to current and future clinical challenges in neuromodulation therapy. Front Neurol. 2020;11:597451.

24. Goodman WK, Alterman RL. Deep brain stimulation for intractable psychiatric disorders. Annu Rev Med. 2012;63:511–24.

25. Pepper J, Hariz M, Zrinzo L. Deep brain stimulation versus anterior capsulotomy for obsessive-compulsive disorder: a review of the literature. J Neurosurg. 2015;122(5):1028–37.

26. Filkowski MM, Sheth SA. Deep brain stimulation for depression: an emerging indication. Neurosurg Clin N Am. 2019;30(2):243–56.

27. Beuter A. Deep brain stimulation. In: Jaeger D, Jung R, editors. Encyclopedia of computational neuroscience. New York: Springer; 2015. p. 968.

28. Clark GM. The multichannel cochlear implant for severe-to-profound hearing loss. Nat Med. 2013;19(10):1236–9.

29. El Boghdady N, Kegel A, Lai WK, Dillier N. A neural-based vocoder implementation for evaluating cochlear implant coding strategies. Hear Res. 2016;333:136–49.

30. Reiss LAJ. Cochlear implants and other inner ear prostheses: today and tomorrow. Curr Opin Physio. 2020;18:49–55.

31. Luts H, Eneman K, Wouters J, Schulte M, Vormann M, Buechler M, et al. Multicenter evaluation of signal enhancement algorithms for hearing aids. J Acoust Soc Am. 2010;127(3):1491–505.

32. Disorders N-NIoDaOC. Quick statistics about hearing 2016 https://www.nidcd.nih.gov/health/statistics/quick-statistics-hearing.

33. Sharma SD, Cushing SL, Papsin BC, Gordon KA. Hearing and speech benefits of cochlear implantation in children: a review of the literature. Int J Pediatr Otorhinolaryngol. 2020;133:109984.

34. Chuang AT, Margo CE, Greenberg PB. Retinal implants: a systematic review. Br J Ophthalmol. 2014;98(7):852–6.

35. Mills JO, Jalil A, Stanga PE. Electronic retinal implants and artificial vision: journey and present. Eye. 2017;31(10):1383–98.

36. Mirochnik RM, Pezaris JS. Contemporary approaches to visual prostheses. Mil Med Res. 2019;6 (1):19.

37. Ayton LN, Barnes N, Dagnelie G, Fujikado T, Goetz G, Hornig R, et al. An update on retinal prostheses. Clin Neurophysiol. 2020;131(6):1383–98.

38. Najarpour Foroushani A, Pack CC, Sawan M. Cortical visual prostheses: from microstimulation to functional percept. J Neural Eng. 2018;15(2):021005.

39. Niketeghad S, Pouratian N. Brain machine interfaces for vision restoration: the current state of cortical visual prosthetics. Neurotherapeutics. 2019;16(1):134–43.

40. Ayton LN, Blamey PJ, Guymer RH, Luu CD, Nayagam DA, Sinclair NC, et al. First-in-human trial of a novel suprachoroidal retinal prosthesis. PLoS One. 2014;9(12):e115239.

41. Zrenner E, Bartz-Schmidt KU, Benav H, Besch D, Bruckmann A, Gabel V-P, et al. Subretinal electronic chips allow blind patients to read letters and combine them to words. Proc Biol Sci. 2011;278(1711):1489–97.

42. Stingl K, Bartz-Schmidt KU, Besch D, Braun A, Bruckmann A, Gekeler F, et al. Artificial vision with wirelessly powered subretinal electronic implant alpha-IMS. Proc Biol Sci. 2013;280 (1757):20130077.

43. Palanker D, Le Mer Y, Mohand-Said S, Muqit M, Sahel JA. Photovoltaic restoration of central vision in atrophic age-related macular degeneration. Ophthalmology. 2020;127(8):1097–104.

44. Rothermel A, Kaim H, Gambach S, Schuetz H, Moll S, Steinhoff R, et al.. Subretinal stimulation chip set with 3025 electrodes, spatial peaking filter, illumination adaptation and implant lifetime optimization. In: 42nd annual international conference of the IEEE engineering in medicine & biology society (EMBC); 2020 20–24 July 2020

45. Ho AC, Humayun MS, Dorn JD, da Cruz L, Dagnelie G, Handa J, et al. Long-term results from an epiretinal prosthesis to restore sight to the blind. Ophthalmology. 2015;122(8):1547–54.

46. Humayun MS, Dorn JD, da Cruz L, Dagnelie G, Sahel J-A, Stanga PE, et al. Interim results from the international trial of second sight's visual prosthesis. Ophthalmology. 2012;119(4):779–88.

47. Hornig R, Dapper M, Le Joliff E, Hill R, Ishaque K, Posch C, et al. Pixium vision: first clinical results and innovative developments. In: Gabel VP, editor. Artificial vision: a clinical guide. Cham: Springer; 2017. p. 99–113.

48. Ohta J, Tokuda T, Kagawa K, Sugitani S, Taniyama M, Uehara A, et al. Laboratory investigation of microelectronics-based stimulators for large-scale suprachoroidal transretinal stimulation (STS). J Neural Eng. 2007;4(1):S85–91.

49. Wang V, Kuriyan AE. Optoelectronic devices for vision restoration. Current Ophthalmology Reports. 2020;8(2):69–77.

50. Daschner R, Rothermel A, Rudorf R, Rudorf S, Stett A. Functionality and performance of the subretinal implant chip alpha AMS. Sensors and Materials. 2018;30(2):179–92.
51. Gabel VP, editor. Artificial vision. 1st ed. Cham: Springer; 2017. p. 232.
52. Stett A, Mai A, Herrmann T. Retinal charge sensitivity and spatial discrimination obtainable by subretinal implants: key lessons learned from isolated chicken retina. J Neural Eng. 2007;4(1): S7–16.
53. Eckhorn R, Stett A, Schanze T, Gekeler F, Schwahn H, Zrenner E, et al. Physiologische Funktionsprüfungen von Retinaimplantaten an Tiermodellen. Ophthalmologe. 2001;98 (4):369–75.
54. Kanda H, Morimoto T, Fujikado T, Tano Y, Fukuda Y, Sawai H. Electrophysiological studies of the feasibility of suprachoroidal-transretinal stimulation for artificial vision in normal and RCS rats. Invest Ophthalmol Vis Sci. 2004;45(2):560–6.
55. Zrenner E, Bartz-Schmidt KU, Besch D, Gekeler F, Koitschev A, Sachs HG, et al. The subretinal implant ALPHA: implantation and functional results. In: Gabel VP, editor. Artificial vision: a clinical guide. Cham: Springer; 2017. p. 65–83.
56. Zrenner E, Stett A, Rubow R. The challenge to meet the expectations of patients, ophthalmologists and public health care systems with current retinal prostheses. Investigative Ophthalmology & Visual Science. 2020;61(7):2214.
57. Fernández E, Alfaro A, González-López P. Toward long-term communication with the brain in the blind by Intracortical stimulation: challenges and future prospects. Front Neurosci. 2020;14:681.
58. Lewis PM, Rosenfeld JV. Electrical stimulation of the brain and the development of cortical visual prostheses: an historical perspective. Brain Res. 1630;2016:208–24.
59. Maynard EM, Nordhausen CT, Normann RA. The Utah intracortical electrode array: a recording structure for potential brain-computer interfaces. Electroencephalogr Clin Neurophysiol. 1997;102(3):228–39.
60. Utah electrode array. https://www.blackrockmicro.com/electrode-types/utah-array/.
61. Troyk PR. The intracortical visual prosthesis project. In: Gabel VP, editor. Artificial vision: a clinical guide. Cham: Springer; 2017. p. 203–14.
62. Rush AD, Troyk PR. A power and data link for a wireless-implanted neural recording system. IEEE Trans Biomed Eng. 2012;59(11):3255–62.
63. Rosenfeld JV, Wong YT, Yan E, Szlawski J, Mohan A, Clark JC, et al. Tissue response to a chronically implantable wireless intracortical visual prosthesis (Gennaris array). J Neural Eng. 2020;17(4):046001.
64. Williams DF. On the mechanisms of biocompatibility. Biomaterials. 2008;29(20):2941–53.
65. Hassler C, Boretius T, Stieglitz T. Polymers for neural implants. J Polym Sci B Polym Phys. 2011;49(1):18–33.
66. Petrossians A, Whalen JJ, Weiland JD. Improved electrode material for deep brain stimulation. In: Annual international conference of the IEEE engineering in medicine and biology society IEEE engineering in medicine and biology society annual international conference 2016; pp 1798–801.
67. Pancrazio JJ, Deku F, Ghazavi A, Stiller AM, Rihani R, Frewin CL, et al. Thinking small: progress on microscale neurostimulation technology. Neuromodulation: Journal of the International Neuromodulation Society. 2017;20(8):745–52.
68. Wang K, Frewin CL, Esrafilzadeh D, Yu C, Wang C, Pancrazio JJ, et al. High-performance graphene-fiber-based neural recording microelectrodes. Advanced Materials. 2019;31(15): e1805867.
69. Hämmerle H, Kobuch K, Kohler K, Nisch W, Sachs H, Stelzle M. Biostability of microphotodiode arrays for subretinal implantation. Biomaterials. 2002;23(3):797–804.

70. Lancaster MA, Knoblich JA. Organogenesis in a dish: modeling development and disease using organoid technologies. Science. 2014;345(6194):1247125.

71. Clevers H. Modeling development and disease with organoids. Cell. 2016;165(7):1586–97.

72. Evans M. Discovering pluripotency: 30 years of mouse embryonic stem cells. Nat Rev Mol Cell Biol. 2011;12(10):680–6.

73. Takahashi K, Yamanaka S. Induction of pluripotent stem cells from mouse embryonic and adult fibroblast cultures by defined factors. Cell. 2006;126(4):663–76.

74. Young HE, Black AC Jr. Adult stem cells. Anat Rec A: Discov Mol Cell Evol Biol. 2004;276 (1):75–102.

75. Young HE, Duplaa C, Romero-Ramos M, Chesselet M-F, Vourc'h P, Yost MJ, et al. Adult reserve stem cells and their potential for tissue engineering. Cell Biochem Biophys. 2004;40 (1):1–80.

76. Hofer M, Lutolf MP. Engineering organoids. Nature Reviews Materials. 2021;3:18.

77. Liu H, Wang Y, Wang H, Zhao M, Tao T, Zhang X, et al. A droplet microfluidic system to fabricate hybrid capsules enabling stem cell organoid engineering. Advanced Science. 2020;7 (11):1903739.

78. Li Y, Tang P, Cai S, Peng J, Hua G. Organoid based personalized medicine: from bench to bedside. Cell Regeneration. 2020;9(1):21.

79. Wei Y, Zhang C, Fan G, Meng L. Organoids as novel models for embryo implantation study. Reproductive Sciences. 2021;28(6):1637–43.

80. Dutta D, Heo I, Clevers H. Disease modeling in stem cell-derived 3D organoid systems. Trends Mol Med. 2017;23(5):393–410.

81. Corrò C, Novellasdemunt L, Li VSW. A brief history of organoids. Am J Phys Cell Phys. 2020;319(1):C151–C65.

82. van den Berg A, Mummery CL, Passier R, van der Meer AD. Personalised organs-on-chips: functional testing for precision medicine. Lab Chip. 2019;19(2):198–205.

83. Kieninger J, Weltin A, Flamm H, Urban GA. Microsensor systems for cell metabolism – from 2D culture to organ-on-chip. Lab Chip. 2018;18(9):1274–91.

84. Kratz SRA, Höll G, Schuller P, Ertl P, Rothbauer M. Latest trends in biosensing for microphysiological organs-on-a-chip and body-on-a-chip systems. Biosensors. 2019;9:3.

85. Ahadian S, Civitarese R, Bannerman D, Mohammadi MH, Lu R, Wang E, et al. Organ-on-a-chip platforms: a convergence of advanced materials, cells, and microscale technologies. Adv Healthc Mater. 2018;7(2):1700506.

86. Picollet-D'hahan N, Zuchowska A, Lemeunier I, Le Gac S. Multiorgan-on-a-chip: a systemic approach to model and decipher inter-organ communication. Trends Biotechnol. 2021;

87. Ingber DE. Is it time for reviewer 3 to request human organ Chip experiments instead of animal validation studies? Advanced Science. 2020;7(22):2002030.

88. Polini A, Moroni L. The convergence of high-tech emerging technologies into the next stage of organ-on-a-chips. Biomaterials and Biosystems. 2021;1:100012.

89. Del Sol A, Thiesen HJ, Imitola J, Carazo Salas RE. Big-data-driven stem cell science and tissue engineering: Vision and unique opportunities. Cell Stem Cell. 2017;20(2):157–60.

90. Kleensang A, Maertens A, Hartung T. From big data to predictive analysis from in vitro systems. In: Murphy SV, Atala A, editors. Regenerative medicine technology. Gene and cell therapy. 1st ed. New York: CRC Press; 2017.

91. Zhu H. Big data and artificial intelligence modeling for drug discovery. Annu Rev Pharmacol Toxicol. 2020;60(1):573–89.

Appendix

Suggested Books for In-Depth Learning

General Overview

Marks et al.: Handbook of Biosensors and Biochips, Wiley, ISBN 978-0-470-01905-4
Morris and Langari: Measurement and Instrumentation, Academic Press, ISBN 978-0-12-800884-3

Chemistry

Goldberg: Fundamentals of Chemistry, McGraw-Hill, ISBN 978-0-073-22104-5
Olmsted and Williams: Chemistry, Wiley, ISBN 978-0-471-47811-9

Biology

Alberts et al.: Molecular Biology of the Cell, Norton & Company, ISBN 978-0-815-34464-3
Wilson and Walker: Principles and Techniques of Biochemistry and Molecular Biology, Cambridge University Press, ISBN 978-0-521-51635-8

Electrochemistry

Bard and Faulkner: Electrochemical Methods, Wiley, ISBN 978-0-471-04372-0
Wang: Analytical Electrochemistry, Wiley, ISBN 978-0-471-79029-7

© The Author(s), under exclusive license to Springer Nature Switzerland AG 2021   323
A. Pasquarelli, *Biosensors and Biochips*, Learning Materials in Biosciences,
https://doi.org/10.1007/978-3-030-76469-2

Microfluidics

Song et al.: Microfluidics, Wiley, ISBN 978-3-527-80062-9
   Tabeling: Introduction to Microfluidics, Oxford University Press, ISBN 978–0–19–856864–3

Material Science and Technology

Cardarelli: Materials Handbook, Springer, ISBN 978-3-319-38925-7
   Frey and Khan: Handbook of Thin-Film Technology, Springer, ISBN 978-3-642-05430-3
   Kalia: Polymeric Hydrogels as Smart Biomaterials, Springer, ISBN 978-3-319-25322-0

Neuroscience

Luo: Principles of Neurobiology, Garland Science, ISBN 978-0-8153-4494-0
   Purves et al.: Neuroscience, Sinauer Associates, ISBN 978-1-605-35380-7

Environmental Monitoring

Artiola et al.: Environmental Monitoring and Characterization, Academic Press, ISBN 978-0-120-64477-3
   Patnaik: Handbook of Environmental Analysis, CRC Press, ISBN 978-1-4987-4561-1

Food Safety

Ortega: Foodborne Parasites, Springer, ISBN 978-0387-30068-9
   Weidenbörner: Mycotoxins in Foodstuffs, Springer, ISBN 978-0-387-73688-4

Security

Nikolelis and Nikoleli: Biosensors for Security and Bioterrorism Applications, Springer, ISBN 978-3-319-28926-7

Enzymes

Bisswanger: Enzyme Kinetics, Principles and Methods, Wiley, ISBN 978-3-527-80647-8

Punekar: Enzymes: Catalysis, Kinetics and Mechanisms, Springer, ISBN 978-981-13-0785-0

Immune System

Delves et al.: Roitt's Essential Immunology, Wiley, ISBN 978-1-118-41606-8

Owen et al.: Kuby Immunology, Macmillan Education, ISBN 978-14641-3784-6

SPR

Homola: Surface Plasmon Resonance Based Sensors, Springer, ISBN 978-3-540-33919-9

Maier: Plasmonics, fundamentals and applications, Springer, ISBN 978-0-387-33150-8

QCM

Janshoff and Steinem: Piezoelectric Sensors, Springer, ISBN 978-3-540-36568-6

Biochips

Appasani: Bioarrays, Springer, ISBN 978-1-58829-476-0

Hierlemann: Integrated Chemical Microsensor Systems in CMOS Technology, Springer, ISBN 978-3-540-27372-1

Pevsner: Bioinformatics and Functional Genomics, Wiley, ISBN 978-1-118-58178-0

Electrophysiology

Dallas and Bell: Patch Clamp Electrophysiology, Springer, ISBN 978-1-0716-0817-3

Fermini and Priest: Ion Channels, Springer, ISBN 978-3-540-79728-9

Taketani and Baudry: Advances in Network Electrophysiology, Springer, ISBN 978-0387-25857-7

Implants

Carrara and Iniewski: Handbook of Bioelectronics, Cambridge University Press, ISBN 978-1-107-04083-0

Katz: Implantable Bioelectronics, Wiley, ISBN 978-3-527-67317-9

Kilgore: Implantable Neuroprostheses for Restoring Function, Woodhead Publishing, ISBN: 978-1-78242-101-6

Humayun and Olmos de Koo: Retinal Prosthesis, Springer, ISBN 978-3-319-67260-1

Organoids and Organs-on-a-Chip

Murphy and Atala: Regenerative Medicine Technology, CRC Press, ISBN 978-1-4987-1191-3

Index

A

Absorption, 78, 90, 91, 97, 104, 107, 110–113, 122, 123, 145, 150, 158, 269, 270, 286
Action potential, 20, 275, 276, 279–282, 287, 292–295, 318
Adsorption, entrapment and covalent binding, 162
Affinity receptors, 22
Amplification, 58, 84, 86, 88, 143, 195–201, 220, 226–228, 230, 231, 234, 239, 251, 253, 262, 286
Antibody, 5, 9, 14, 16, 20–23, 31, 32, 42, 43, 78–85, 87, 88, 92, 93, 98–101, 113, 117, 119–121, 128, 133, 134, 137, 139–141, 152, 157, 162, 164, 170, 171, 176, 262, 263, 268, 270, 272, 273, 310
Antigen, 5, 20–22, 78–81, 83, 87, 92, 93, 99, 119, 128, 140, 141, 263, 265, 268, 274
Applications, 2–4, 6, 9–17, 21–23, 42, 43, 46, 57, 58, 60, 70, 71, 74, 107, 108, 111, 131, 134, 136, 153, 162, 165, 186, 190–192, 203, 217, 221, 270, 284, 292, 296, 297, 300, 311, 313, 315, 318
Aptamer, 9, 20, 23, 82–85, 101
Arraying, 143, 181, 183, 206

B

Bioreceptor, 2–6, 16, 17, 20–43, 46, 70, 73, 78, 111, 143, 152, 161–183
Bridge amplification, 230, 231
Bulk acoustic waves (BAW), 146–150, 153, 156–158

C

Calorimetric detection, 56–58

Cantilevers, 78, 143–146, 149, 158
Catalysis, 5, 20, 46–54, 75
Catalytic receptors, 72
Chemical bonds, 26–31, 42, 43
Clustering, 217–222, 261, 264, 265, 267

D

Definition, 2, 107, 140, 228
Dideoxynucleotide, 226, 237, 251
DNA, 5, 9, 10, 20, 22–23, 42, 43, 78, 82–84, 152, 162, 172, 186–193, 195–198, 200–210, 212, 214, 220–222, 226–228, 230, 232–235, 237, 238, 241–243, 245–253, 262, 264, 267, 286, 287, 295

E

Electrochemical detection methods, 75
Electropherogram, 227, 251
Emulsion PCR, 232, 234
Epitope, 22, 78, 79, 81, 99, 140, 141
Evanescent waves, 116, 117, 123, 156

F

Function recovery, 70

G

Gel electrophoresis, 198, 200, 201, 226–228, 232, 251, 259–261
Genes
    expressions, 189, 191, 192, 203, 207, 212, 214, 216, 218, 221, 226, 242, 287, 313
Genome assemblies, 243

Printed in the United States
by Baker & Taylor Publisher Services